Multilevel Inverters

Multilevel Inverters

Conventional and Emerging Topologies and their Control

Krishna Kumar Gupta, PhD, MIEEE

Pallavee Bhatnagar, PhD, MIEEE

ELSEVIER

ACADEMIC PRESS

An imprint of Elsevier

Academic Press is an imprint of Elsevier
125 London Wall, London EC2Y 5AS, United Kingdom
525 B Street, Suite 1800, San Diego, CA 92101-4495, United States
50 Hampshire Street, 5th Floor, Cambridge, MA 02139, United States
The Boulevard, Langford Lane, Kidlington, Oxford OX5 1GB, United Kingdom

British Library Cataloguing-in-Publication Data
A catalogue record for this book is available from the British Library

Library of Congress Cataloging-in-Publication Data
A catalog record for this book is available from the Library of Congress

ISBN: 978-0-12-812448-2

For Information on all Academic Press publications
visit our website at https://www.elsevier.com/books-and-journals

Publisher: Joe Hayton
Acquisition Editor: Lisa Reading
Editorial Project Manager: Lindsay Lawrence
Production Project Manager: Vijayaraj Purushothaman
Cover Designer: Mark Rogers

Typeset by MPS Limited, Chennai, India

Dedication

Dedicated to Lord Shri Krishna

Contents

Foreword

(IEEE William E. Newell Power Electronics Award, 2014, Formerly, Editor-in-Chief, IEEE Transactions on Power Electronics)

Power electronics is a key technology in modern society that brings better sustainability as well as better effectiveness in manufacturing. Today, renewable sources-based power generation is booming and has power electronics as the key technology that enables grid integration. We also see strong indications for maturation of the electrical car industry within the next decade. Further, the way to save energy (e.g., in heating, ventilation and air-conditioning) is to use variable-speed drives composed of power electronics technology.

Therefore, "Multilevel Inverters: Conventional and Emerging Topologies and their Control" is an important contribution to this field of power electronics, as multilevel converters can be applied within any power range—both at low power (as in solar energy systems), very high power when used for power converters in wind turbines, as well in as very large-scale power transmission for the grid. New theories need to be developed, practical hardware implementations need to be innovated and, of course, demonstrations and validation need to show that the technology really works. Multilevel converters have also seen many applications as they may offer better reliability due to their possible redundant design enabling the converter to operate if one module/leg/switch fails. Future challenges are for better models including small signal models, as well as serious analysis of the failure operation.

I wish you an enjoyable reading!

By Prof. Frede Blaabjerg, Fellow, IEEE
Professor and VILLUM Investigator, Aalborg University, Aalborg, Denmark

About the Authors

Krishna Kumar Gupta is an avid teacher, researcher, consultant and an author. Dr. Gupta is currently serving as an Associate Professor, lecturing on Power Electronics, Control Systems and Circuit Theory. In the year 2015, Dr. Gupta was conferred upon with Young Scientist Award by the Government of Madhya Pradesh, India for his research on inverters. In 2016, he has been awarded by the Confederation of Indian Industries (CII) for his contribution in teaching.

Before taking up research on inverters, Dr. Gupta had varying tenures in Bharat Heavy Electricals Limited (BHEL), Tata Consultancy Services (TCS) and National Institute of Technology (NIT) Raipur, India. Post PhD, he was offered a research position at Nanyang Technological University (NTU), Singapore but he was more keen to work on development of technologies for PV based power generation for rural India.

Dr. Gupta is a Member, Institute of Electrical and Electronics Engineers (IEEE) and has authored several technical papers. He obtained his B.Tech., M.Tech. and Ph.D. degrees from Maulana Azad National Institute of Technology (MANIT), Bhopal, India. He also serves as an active reviewer for many journals of international repute.

An entrepreneur, professional engineer, researcher and academician, **Pallavee Bhatnagar** is a Ph.D. in Power Electronics from Maulana Azad National Institute of Technology (MANIT), Bhopal. Dr. Bhatnagar is currently serving as Principal at Sagar Institute of Science and Technology, Bhopal, India. She concurrently owns and runs a technological enterprise which manufactures electrical equipment. Prior to these, she has worked as an Engineer at National Thermal Power Corporation Limited (NTPC) and All India Radio (AIR). She also works for the social upliftment of women through science and technology and has been honored by the Member of Parliament, Bhopal for the same.

Dr. Bhatnagar is a passionate teacher too. Her subjects of special interests are Electrical Machines, Power Electronics and Control Systems. In addition, she is also heading a research on development of efficient power electronics drivers for HBLEDs under the National Perspective Plan (NPP) funded by the Ministry of Power, Government of India through Central Power Research

Institute (CPRI), Bangalore, India. Dr. Bhatnagar is a Member, Institute of Electrical and Electronics Engineers (IEEE) and Life Member, Institutions of Engineers India (IEI). She has several research papers to her credit.

Preface

No power on Earth can stop an idea whose time has come.

–Victor Hugo

This quote is certainly true for the technology that we call power electronics. In the past few decades, power electronics has changed the face of electrical engineering. It has changed the way we generate, transmit, distribute, and utilize electric power. Power electronics is also an enabling technology for smart grids, renewable energy-based power generation, electric propulsion, cell (mobile) phones, self-driving cars, etc. In this book, we start with a discussion on the philosophy of power electronics. Of course, the main discussion is on multilevel inverters (MLIs), but we sincerely feel that an appreciation of the idea of power electronics does lead to a better understanding of MLIs.

In recent years, MLIs have gained immense importance and popularity in the industrial and research communities. MLIs have been introduced in many universities, even at the undergraduate level. This book is in response to the need of teachers, researchers, students, and practising engineers due to the lack of organized and comprehensive material on MLIs.

Writing this book was very challenging for us. This was mainly because there are many terminologies related to MLIs which are still not "settled." In the currently available literature on MLIs, while using the same term, various authors imply different meanings! Unlike many other technical topics in which an author may refer to other books to consolidate her thoughts, no such works are currently available when it comes to the discussion on MLIs. And hence, we were heavily dependent on our research experience, in addition to references in the form of numerous technical papers.

The table of contents for this book may give the impression that this is a presentation of a "specific research" and its findings. But no! Some specific research findings are indeed presented, but in a manner that will strengthen your own understanding of MLIs and pave a way for empowering you toward your own contribution in this field.

We hope that this book will enhance your understanding of power electronics in general and MLIs in particular. Please feel free to contact us for feedback, discussions, and criticism!

Dr. Krishna Kumar Gupta and Dr. Pallavee Bhatnagar

Acknowledgments

We are deeply indebted to Dr. P.V. Krishnan, former Professor, Department of Applied Mechanics, Indian Institute of Technology Delhi, India, for accepting us as students and for guiding us in every possible way. We are also grateful to Mrs. and Mr. Muneesh Garg, Indian Revenue Service, for their unwavering support.

We acknowledge the guidance that we received from Dr. Sanjeet Dwivedi, Senior R&D Control Engineer, Global Design Centre, Danfoss Power Electronics, at various stages of this project.

While working on this book, we have been fortunate to have been associated with Ms. Lisa Reading, Senior Acquisition Editor (Energy) and Ms. Lindsay C. Lawrence, Editorial Project Manager. They have been extremely patient with us and have been fully supportive.

We are extremely grateful to our teachers Dr. R.K. Nema, Professor, Maulana Azad National Institute of Technology and Dr. Shailendra Jain, Director, Sant Longowal Institute of Engineering and Technology, Punjab.

Krishna Kumar Gupta is grateful to his parents, Mr. Jai Gopal and Mrs. Kanta Gupta, for their love and support. He is also grateful to his wife Vandana for her love, care, and patience. Mr. Amit Jaiswal and Ms. Jyoti Jaiswal have been a source of strength in his life. His friend Mr. Jitendra Ambulkar helped him in many ways to make this book a reality.

Pallavee Bhatnagar deeply acknowledges the unwavering support and love that she has received from her parents Mr. Om Prakash and Mrs. Usha Shrivastava. Her husband Vivek has been a source of strength and wisdom to her. She is also grateful to her loving daughter Sunidhi Bhatnagar, who has been a fountain of happiness in her life. She is also grateful to her sister Ms. Pragya Nigam for her constant support.

Dr. Krishna Kumar Gupta and Dr. Pallavee Bhatnagar

Chapter 1

Basics of Inverters

1.1 INTRODUCTION

While the demand for electrical energy continues to grow throughout the world, it is estimated that 100% of electrical energy will flow through power electronics, especially in developed countries [1]. It is anticipated that the role of power electronics in this era will be as crucial as that of computers and communication technologies. Power electronics has changed the face of electrical engineering. Applications of power electronics include regulated power supplies, uninterruptible power supplies (UPS), electrochemical processes, heating and lighting control, welding, reactive power compensation, flexible AC transmission systems, active filters, energy storage, motor drives, and so on. Of all power electronics converters, DC to AC converters (i.e. "inverters") are most widely employed in the modern set-up of electric power generation, transmission, distribution, utilization, and protection.

This chapter deals with the fundamentals of inverters, however before that, there is a discussion on the power electronics technology. Throughout this book, concepts are discussed mainly in the context of voltage source inverters because the voltage-fed class of inverters is becoming universal, replacing the current-fed class [1]. While you might already have a good understanding of inverters, we would still recommend you read this chapter. This chapter serves as the base for building up a case for multilevel inverters. It will help you understand and appreciate the concept behind multilevel inverters from various angles. Moreover, it will help you understand the popular terminology for multilevel inverters in research publications.

In addition to the basics of the so-called H-bridge inverter, we also discuss the philosophy behind waveform analysis and filtering needs. This discussion sets up the scene for the pulse-width modulation (PWM) technique. In the final part, power switch requirements are discussed for inverters in general. Some more specific cases in the context of multilevel inverters will be taken up in subsequent chapters.

1.2 POWER ELECTRONICS AS A TECHNOLOGY

The purpose of this section is to understand the philosophy of power electronics in a simplified manner. Once this philosophy is understood, it will be

Multilevel Inverters. DOI: https://doi.org/10.1016/B978-0-12-812448-2.00001-5

extremely easy to introduce and analyze various concepts related to inverters at this stage and those related to multilevel inverters at later stages.

Before defining power electronics, let us consider a hypothetical problem (without considering practical application, as of now): you have a resistance of 10 Ω which is to be supplied with a voltage of 8 V, while you have a battery of 12 V at your disposal. What do you do? Well, you visualize the problem as shown in Fig. 1.1. You recognize that the battery (the "source") is, of course, to be connected to the resistance (the "load"), but since the required voltage and the available voltage do not match, you need to do "something" in between! And that "something" should "step-down" the voltage from 12 to 8 V.

Conventional wisdom tells you to connect another resistance of 5 Ω "in between" and voila! You do get the required voltage at the load terminals, as shown in Fig. 1.2. But there is a catch. The 5 Ω resistance dissipates a power of 3.2 W, while the source delivers a power of 6.4 W. So, the power loss is 50% of delivered power!

So you now look for another solution which fulfills two requirements: it steps down voltage and it does so without any power loss (ideally). You connect a "switch *S*" in between the source and the load, as shown in Fig. 1.3A. And, you repeatedly operate the switch *S* in ON and OFF states. When *S* is

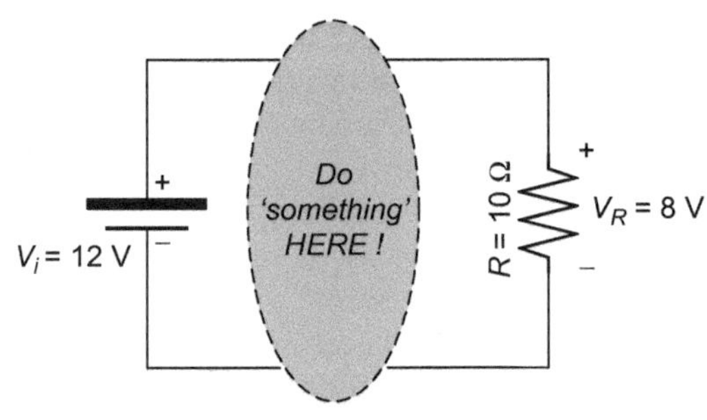

FIGURE 1.1 A hypothetical problem where a resistance is to be supplied with a voltage of 8 V while a source of 12 V is available.

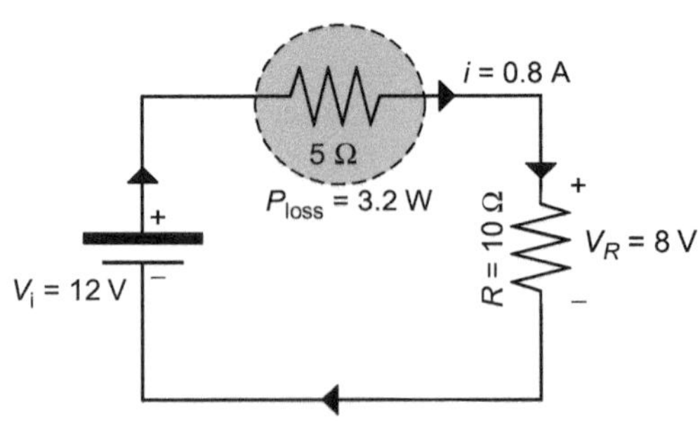

FIGURE 1.2 While putting another resistance in series helps to obtain the desired load voltage, it also leads to power loss.

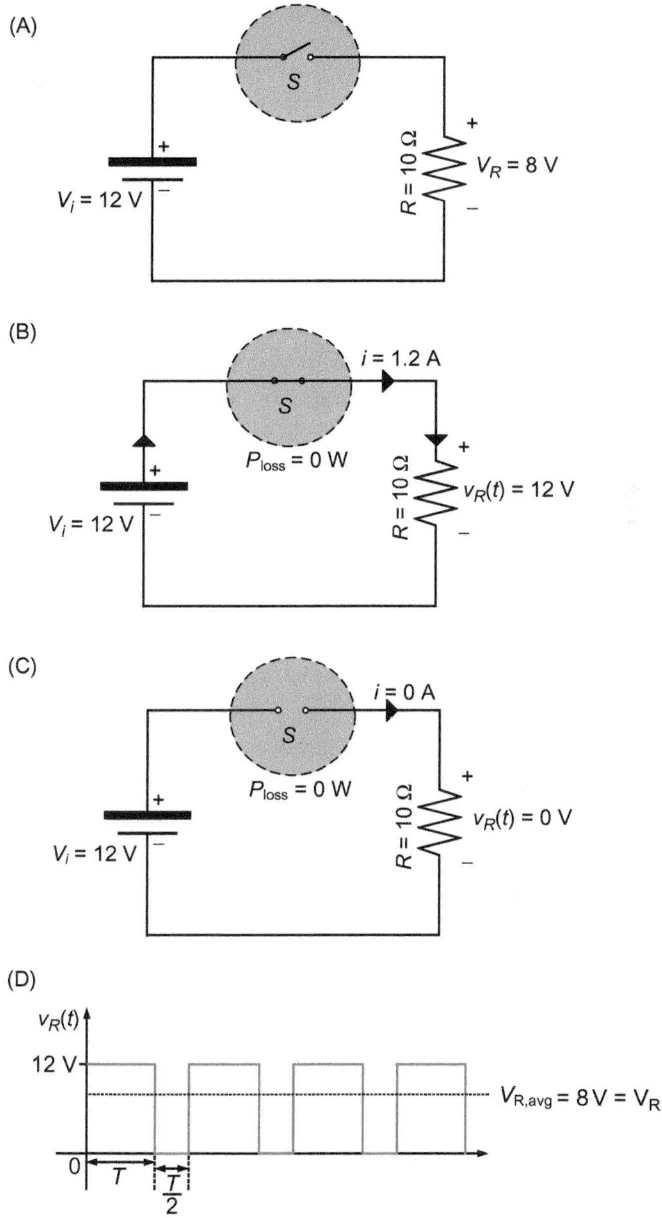

FIGURE 1.3 (A) Putting a switch between the source and the load to control the load voltage; (B) when the switch is ON; (C) when the switch is OFF; (D) voltage waveform across the load.

ON, the instantaneous voltage $v_R(t)$ across the load is 12 V and when S is OFF, $v_R(t)$ is 0 V. These modes are shown in Fig. 1.3B and C, respectively. Your logic of doing so is simple: you are controlling the "average" voltage across the load. That is, if you keep S ON for a time duration T, then next you keep it OFF for a duration $T/2$ and you keep repeating it. You get a voltage waveform across the load resistance as shown in Fig. 1.3D and you calculate the average value of voltage to be 8 V (= V_R)! Also, when ON, the switch carries a current of 1.2 A but the voltage across it is zero (ideally) and hence power loss is zero. Similarly, when OFF, the voltage across the switch is 12 V but the current through it is zero and hence, again, power loss is zero! And therefore, the arrangement works well!

The arrangement indeed works well but you are dissatisfied with one thing: the voltage across the load is "fluctuating." So you decide to remove the ripples by placing a capacitor across the load, as shown in Fig. 1.4A. But then, when the capacitor C is completely uncharged initially and you operate S to the ON position, you see that the capacitor acts as a short-circuit, thereby directly short-circuiting the source and damaging the circuit. In order to prevent this, you simply put an inductor L in series with the capacitor C. That way, because the inductor will initially act as an open-circuit, switching ON S when capacitor C is uncharged, will not lead to the short-circuiting of the source. The configuration is shown in Fig. 1.4B. But then, you recall the property of an inductor: it opposes instantaneous change in current and "produces" a voltage governed by the first derivative of the current through it. So, when the switch S in the circuit is carrying current and is turned OFF, a large voltage is produce which can breakdown the switch S itself. You recognize the need to have a path in the circuit such that when the switch S is turned OFF, the current through the inductor finds a way and does not suddenly reduce to zero. And hence you introduce a diode for the purpose, as shown in Fig. 1.4C.

The final circuit shown in Fig. 1.4C can be immediately recognized as a "buck" topology.

From this discussion, we intend to describe the definition of "power electronics." *Power electronics is a technology which deals with the conversion and control of electrical power with the help of power semiconductor devices that operate in switching mode, and therefore, the efficiency of the power electronics apparatus may approach as high as 98–99%* [1]. The underlined words are of utmost importance. With respect to the example here, we see that we are dealing with a conversion process: a fixed DC voltage of 12 V is being converted to a regulated voltage of 8 V (i.e., it can be changed to any other average value with proper switching). So, the process here is *DC to DC conversion*. Similar conversions can be: *DC to AC, AC to DC*, and *AC to AC*. We also observe that the *switch* helps us to do the same. When it is turned ON, it connects the source and the load and when OFF, it disconnects them. So basically, with this switching effort, we are controlling the electrical power in the circuit. The switch is implemented with a power semiconductor device. Thus, operation of power semiconductor devices as *switches*

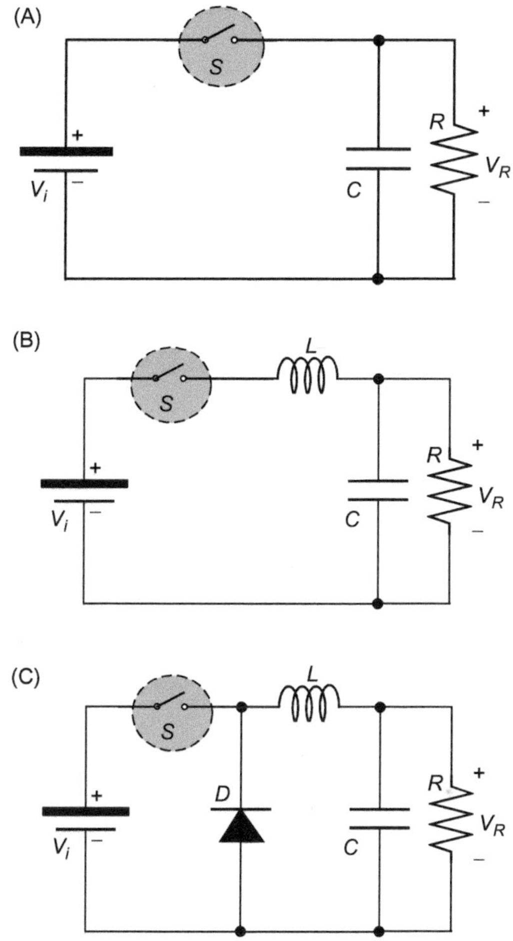

FIGURE 1.4 Description and derivation of "buck" topology.

is what differentiates power electronics from *linear electronics* (in which a semiconductor device is operated in its linear region). And what about the inductor and capacitor in the circuit here? Well, they act as energy storage components and they enable energy continuity when the energy flow is "interrupted" by the switching function.

Hence, operating a semiconductor device as a switch is the underlying philosophy of power electronics and all types of conversions, as mentioned previously, can be carried out with it. Now, the "circuits" which perform the function of "conversion" are popularly referred to as "converters." At this point, a differentiation must be mentioned between a "topology (or a circuit)" and a "converter." In the past, it was customary to refer to the term "converter" as an electrical circuit diagram of the connection of power electronics components; i.e., the so-called abstract "circuit" [2]. This is an

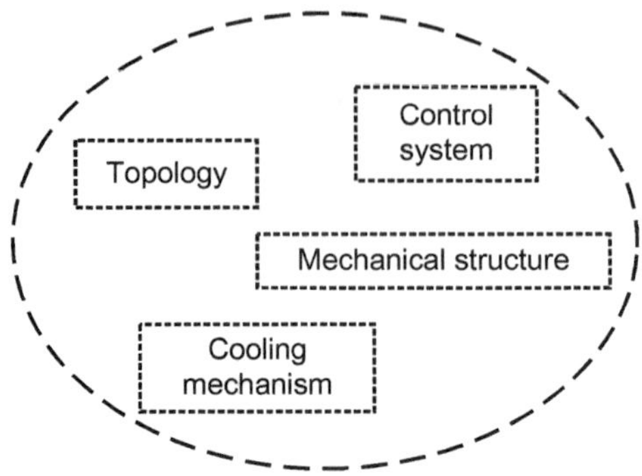

FIGURE 1.5 "Topology" is just a part of the "converter," though the terms are often used interchangeably.

extremely restrictive description. For all practical purposes, a converter will be taken as the total of the equipment between the source and the load that have the objective of the conversion and control of the electromagnetic energy flow between an electric source and the load [2]. The converter has the topology as its part in addition to the thermal and structural components. The overall set of functionalities that a practical converter encompasses is shown in Fig. 1.5. Hence, a topology is a part of the overall scheme of things (i.e., the converter). Although the terms topology and converter are often used interchangeably, the reader should always bear in mind the practicality of actual implementation. It should also be mentioned here that for the very same purpose, there can obviously be numerous topologies.

1.3 CONCEPT OF STATES: EXAMPLE OF AN INVERTER

A DC to AC converter is commonly referred to as an **inverter**. In this and following sections, we take the example of a so-called **H-bridge topology** or **H-bridge inverter** to explain a few concepts which are commonly used in the context of DC-to-AC conversion. There is a DC source represented as a battery of voltage V_{DC}, as shown in Fig. 1.6. This input voltage is commonly referred to as the **DC link voltage**. The higher potential terminal of this source is marked as P (also referred to as the **positive DC rail**) and the lower potential terminal is marked as N (also referred to as the **negative DC rail**). The load is shown with a black box (because the load may be resistive, inductive, or capacitive in nature, but more on that later). The load requires an alternating voltage (typically with a sinusoidal waveform) for its operation. The load terminals are indicated as A and B. The load voltage is marked as $v_o(t)$ and the load current is marked as $i_o(t)$. When the potential at A is higher than at B, then $v_o(t)$ is considered positive, otherwise it is negative. Similarly, when the direction of current is from A to B, then $i_o(t)$ is considered positive, otherwise it is negative. The arrangement also has four

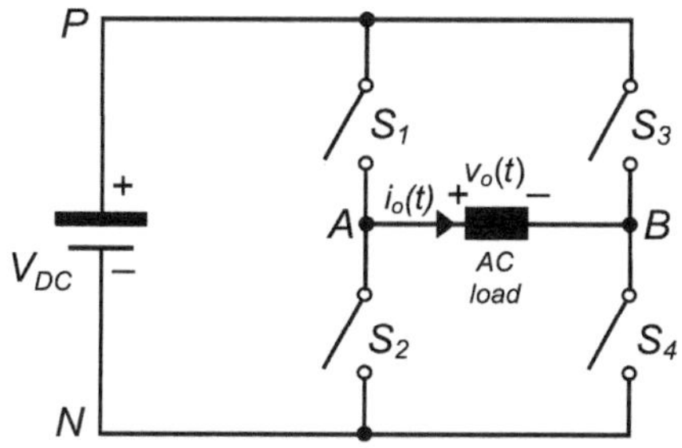

FIGURE 1.6 "Topology" of an H-bridge inverter.

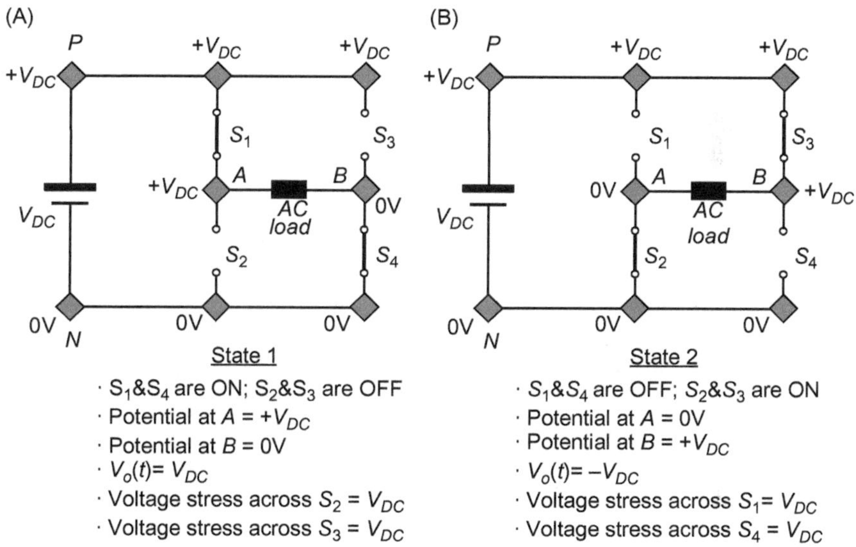

FIGURE 1.7 Switching states for the H-bridge inverter.

switches named S_1, S_2, S_3, and S_4. The power semiconductor devices that can be used as these switches are discussed in Section 1.7.

Now, the operation of this topology for the purpose of DC-to-AC conversion is described with the help of Fig. 1.7. For ease of analysis, point N of the topology is considered to be the reference point with a potential of 0 V. Accordingly, the potential at point P with respect to point N would be $+V_{DC}$. Whenever a switch is turned ON, it is said to be in a **conducting** state and whenever a switched is turned OFF it is said to be in a **blocking** state. The **current rating** of a switch is decided by the highest current that it has to carry in the conducting state, with some additional margin. Whenever a switch is in a blocking state, it has to bear a voltage. This voltage is called **voltage stress** or **standing voltage** for the switch [3]. The **voltage rating** of a switch is decided by the maximum voltage that a switch has to bear.

When switches S_1 and S_4 are switched ON together, keeping switches S_2 and S_3 OFF, then it can be seen from Fig. 1.7A that the load voltage is V_{DC}. When switches S_2 and S_3 are turned ON keeping S_1 and S_4 OFF, then the load voltage is $-V_{DC}$, as can be seen from Fig. 1.7B. These two switching configurations are referred to as **states**. Any valid combination with each switch conducting or blocking is called a state of the topology. Thus, two valid states have been discussed for the H-bridge and these two states lead to load voltage being positive and negative, respectively. Hence if the H-bridge inverter is operated in these two states periodically, then an alternating waveform (a square wave) is obtained across the load, as depicted in Fig. 1.8. We have not considered the current as of now and it will be discussed while addressing the issue of switch configurations in Section 1.7.

The points of significance from this conversion process of DC to AC include:

1. "Switching," which lies at the heart of the philosophy of power electronics, enables this conversion.
2. Though the desired output waveform is a sinusoidal waveform, the obtained waveform is "different." A sinusoidal waveform is a "continuous" waveform and, in principle, it cannot be "synthesized" using switches because a switch has "discrete" states (either ON or OFF). The solution then left is to "imitate" a sinusoidal waveform as much as possible with the help of topology and switching.
3. To obtain an alternating voltage waveform, the states are to be systematically obtained (i.e., state 1 for a specified duration and state 2 for an equal duration, repeatedly). It can be recognized from Fig. 1.7 that state 1 is "actuated" by switching ON S_1 and S_4 simultaneously (and keeping S_2 and S_3 OFF) and state 2 is actuated by switching ON S_2 and S_3 simultaneously (and keeping S_1 and S_3 OFF). In all kinds of conversion processes, such systematic transitions in states are to be carried out based on the desired result. This process is called **modulation**.

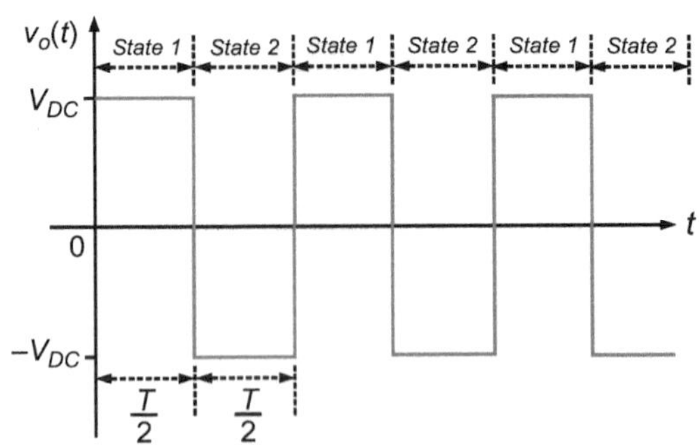

FIGURE 1.8 A square waveform.

4. The peak value of the obtained square wave is equal to the DC link voltage. This peak value is also referred to as the "**operating voltage**."
5. The square waveform so obtained has two distinct values at any point of time: either $+V_{DC}$ or $-V_{DC}$. Such a waveform is called a **two-level waveform**.
6. For all the switches in the H-bridge topology, when they are in the respective blocking state, the voltage stress is equal to the DC link voltage (i.e., V_{DC}). *In H-bridge topology, voltage stress (and hence the minimum voltage rating) is equal to the operating voltage*. This minimum blocking voltage capability required for a switch is also called the **standing voltage**.
7. The H-bridge topology has four switches, each with a minimum blocking voltage requirement (or standing voltage) of V_{DC}. Thus, the total voltage blocking capability required is obtained by addition of the blocking voltage requirement of each switch of a topology. This is called **total standing voltage (TSV)**. Thus, for the H-bridge topology shown in Fig. 1.6,

$$\text{TSV} = V_{DC} + V_{DC} + V_{DC} + V_{DC} = 4V_{DC} \tag{1.1}$$

TSV is also expressed per unit in terms of the operating voltage. In this case,

$$\text{TSV}_{\text{p.u.}} = 4 \tag{1.2}$$

The total standing voltage (or total voltage blocking) for power switches is an important criterion to evaluate the characteristics of an inverter [3]. It gives an idea of the total amount of semiconductor used in the circuit and partly reflects the costs and losses incurred.

8. As far as the effect of the "discrete" nature of the output waveform on the load is concerned, it can be observed that the voltage level periodically changes from $+V_{DC}$ to $-V_{DC}$ in almost no time and again from $-V_{DC}$ to $+V_{DC}$ in almost no time. This rate of change of voltage imposed on the load may cause undesirable effects. For example, in the motor windings, such a voltage may cause leakage of current through the insulation. The arrangement of conductor and insulation acts like a capacitance and the current through a capacitance is dependent on the *rate of change* of voltage across it. That is:

$$i_c(t) = C\frac{dv_c(t)}{dt} \tag{1.3}$$

This rate of change of voltage as seen in the waveform is often expressed as the ***dv/dt* stress on the load**. In this per unit form, the *dv/dt* stress is:

$$\frac{dv}{dt}\text{stress on load} = \frac{1}{(\text{Operating voltage})}\lim_{\Delta t \to 0}\frac{(\text{Change in voltage in a step})}{\Delta t} \tag{1.4}$$

Hence, for the two-level waveform shown in Fig. 1.8, the *dv/dt* stress on the load would be:

$$\frac{dv}{dt}\text{stress in p.u.} = \lim_{\Delta t \to 0} \frac{2}{\Delta t} \tag{1.5}$$

1.4 WAVEFORM ANALYSIS

As discussed in Section 1.3, the desired waveform for an AC load is sinusoidal, whereas the waveform obtained was a square wave. The question arises: how far is the obtained alternating waveform from a sinusoidal waveform? And more importantly, is there a way that the square waveform can be "filtered" to get a sinusoidal waveform?

These questions pertain to "waveform analysis." And they are answered by Fourier's theorem. Before going briefly into this very popular theorem, we first discuss the concept behind it, along with a few definitions. As shown in Fig. 1.9, consider a sinusoidal waveform $y_1(t)$ of a frequency f_o. Let us add to

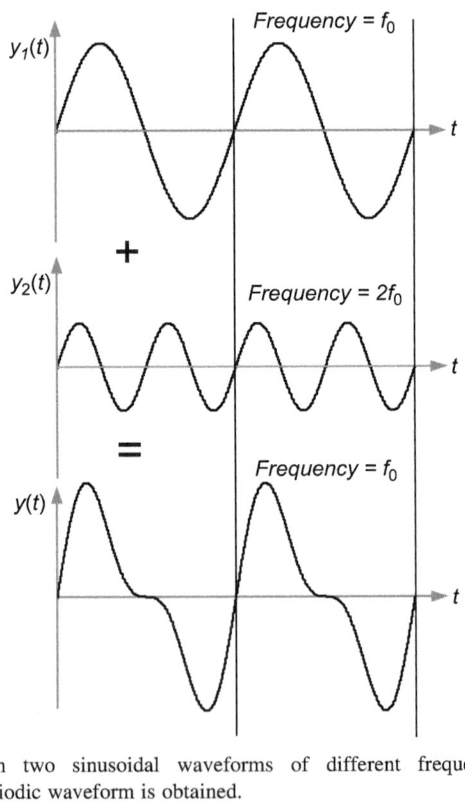

FIGURE 1.9 When two sinusoidal waveforms of different frequencies are added, a nonsinusoidal but periodic waveform is obtained.

it another sinusoidal waveform $y_2(t)$ of frequency $2f_o$ with some arbitrarily chosen amplitude. The result is a waveform $y(t)$ which is nonsinusoidal in nature and has a frequency equal to f_o. This is known as waveform synthesis and in reverse it is called waveform analysis. That is, the nonsinusoidal waveform $y(t)$ can be *broken down* into sinusoidal **components**, or simply sinusoids.

A more elaborate example is shown in Fig. 1.10 in which a periodic but nonsinusoidal waveform $y(t)$ of frequency f_o is broken down to obtain its sinusoidal components. Of these components, one component, $y_1(t)$, has the *same* frequency as that of the original waveform, i.e., a frequency equal to f_o. This component is called the **fundamental component**. All other components have frequencies equal to integral multiples of f_o. In the case of Fig. 1.10, the components of $y(t)$ other than the fundamental component are $y_3(t)$, $y_5(t)$, $y_7(t)$, $y_9(t)$, and $y_{11}(t)$, having respective frequencies of $3f_o$, $5f_o$, $7f_o$, $9f_o$, and $11f_o$. All these components of $y(t)$, excluding the fundamental component $y_1(t)$, are collectively called **harmonics**. Each harmonic has an order associated with it based on the frequency of the component. Thus, $y_3(t)$ is the **third harmonic** and $y_5(t)$, $y_7(t)$, $y_9(t)$, and $y_{11}(t)$ are the fifth, seventh, ninth, and eleventh harmonics, respectively. Representing a nonsinusoidal waveform in terms of its sinusoidal components is a tedious task. For this purpose, a more convenient standard method is to use a **harmonic spectrum**, also shown in Fig. 1.10. A harmonic spectrum has an order of harmonics on the x-axis and respective magnitudes in the y-axis. Generally, the magnitudes of the harmonics are expressed in terms of the percentage of the fundamental component (shown with the order "1"). Thus, for any periodic waveform, the frequency spectrum or the harmonic spectrum shows the components and their relative presence.

The whole of this discussion is an illustration of Fourier's theorem, which states that: "Any arbitrary function can be represented by an infinite series of sinusoids of harmonically related frequencies." That is to say:

$$f(t+T_o) = f(t) = a_o + a_1\cos(\omega_o t) + a_2\cos(2\omega_o t) + \ldots + a_n\cos(n\omega_o t) + \ldots + b_1\sin(\omega_o t) + b_2\sin(2\omega_o t) + \ldots + b_n\sin(n\omega_o t) + \ldots \tag{1.6}$$

Here, $\omega_o = 2\pi f_o = 2\pi/T_o$ and while there is a methodology to calculate the so-called Fourier coefficients a_n and b_n, this is not discussed here. Fourier's series is discussed here to emphasize that the nonsinusoidal waveform obtained from an inverter is analyzed with the help of this theorem, especially in terms of the frequency spectrum. In many softwares, tools are available to obtain the frequency spectrum of a periodic waveform. Such a spectrum for the square waveform, as obtained with MATLAB/Simulink, is shown in Fig. 1.11. It can be seen that the fundamental component is the only desired component as the output of the inverter. All other components (i.e., the harmonics) are undesirable. In other words, these harmonics *corrupt* the fundamental component and render it a nonsinusoidal shape! The extent

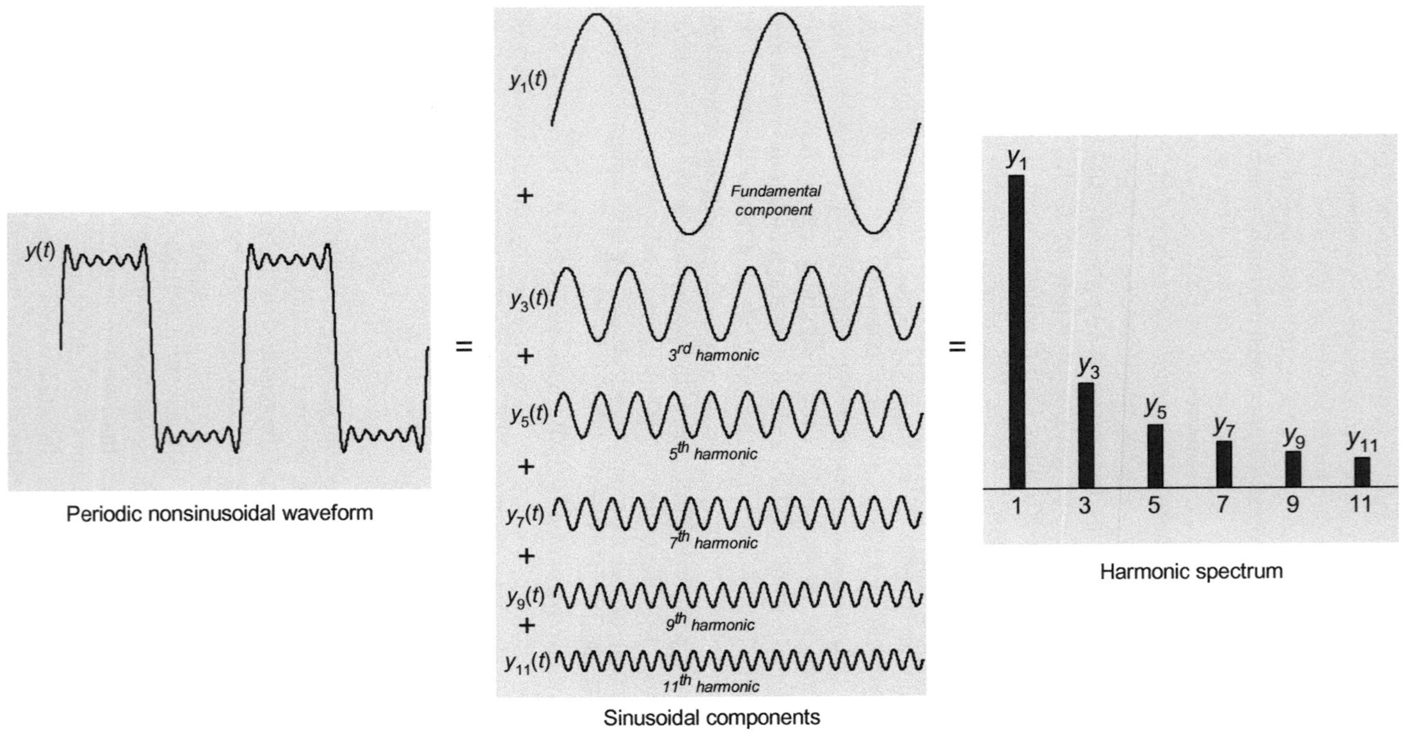

FIGURE 1.10 Concept of analysis of a periodic waveform by breaking it down into sinusoidal components and summarizing it with a harmonic (or frequency) spectrum.

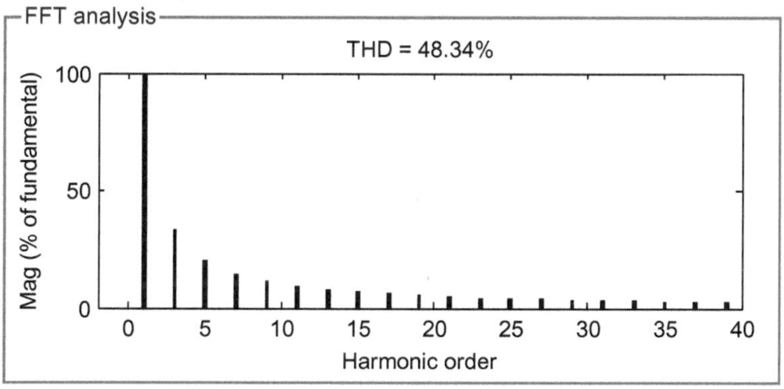

FIGURE 1.11 Frequency spectrum of square-waveform as obtained using MATLAB/Simulink.

of this *distortion* is quantified by a term called **total harmonic distortion (THD)**, expressed as:

$$\text{THD} = \frac{\sqrt{Y_2^2 + Y_3^2 + Y_4^2 + \ldots}}{Y_1} \tag{1.7}$$

where, Y_1, Y_2, Y_3 ... are the root mean square (RMS) values of the frequency components $y_1(t)$, $y_2(t)$, $y_3(t)$ ..., respectively. The greater the THD, the further the waveform is from being sinusoidal.

1.5 FILTERING

What is desired of an inverter is a sinusoidal waveform which cannot be synthesized using power switches. What we obtained in the context of Section 1.3 is a square waveform. The harmonic spectrum of a square waveform is shown in Fig. 1.11 and one wonders if all the harmonics can be *removed* from the spectrum such that only the fundamental component remains. Such *filtering* is actually possible! Recall from the fundamentals that an inductor offers opposition to the flow of sinusoidal current and this opposition (so-called *reactance*) is dependent on the frequency of the current being pushed through the inductor, i.e.,

$$\text{Inductive reactance } X_L = 2\pi f L \tag{1.8}$$

where, L is value of inductance and f is the frequency of the sinusoidal current being *pushed* through the inductor.

Similarly, a capacitor opposes the flow of sinusoidal current through it and this opposition (so-called *capacitive reactance*) is also dependent on the frequency of the sinusoidal current being pushed, i.e.,

$$\text{Capacitive reactance } X_C = 1/(2\pi f C) \tag{1.9}$$

where, C is the value of capacitance and f is the frequency of the sinusoidal current being *pushed* through the capacitor.

A nonsinusoidal waveform is composed of a number of sinusoids (as we now know from Fourier's theorem). Such a waveform, when made to pass through an inductor or a capacitor or a combination of both, has *many* of the sinusoids *filtered* out, leaving only a few. In the context of an inverter, we want the fundamental component (which is of power frequency) to remain and filter out all other components. This concept is shown in Fig. 1.12 where it is shown how an inductor, placed in series, offers large impedance to sinusoids of large frequencies and allows the sinusoids of low frequencies to *pass*. Similarly, when placed in parallel, a capacitor offers large impedance to the components of large frequencies and lets the lower frequency component pass, as shown in Fig. 1.13. Thus, an inductor in series and a capacitor in parallel are used in combination to form a **low-pass LC-filter** as shown in Fig. 1.14.

So, can we get a pure sinusoidal waveform from an inverter? The answer is: not directly, we need to use a filter. Now, we come to a more practical and important point which takes us to the point of the concept of high-frequency PWM. If we have a look at the frequency spectrum of a square waveform, as shown in Fig. 1.11, we see that the lower-order harmonics (of the orders third, fifth, etc.) are prominently present. Filtering out the lower-order harmonics requires bulky inductors and capacitors. For example, let us

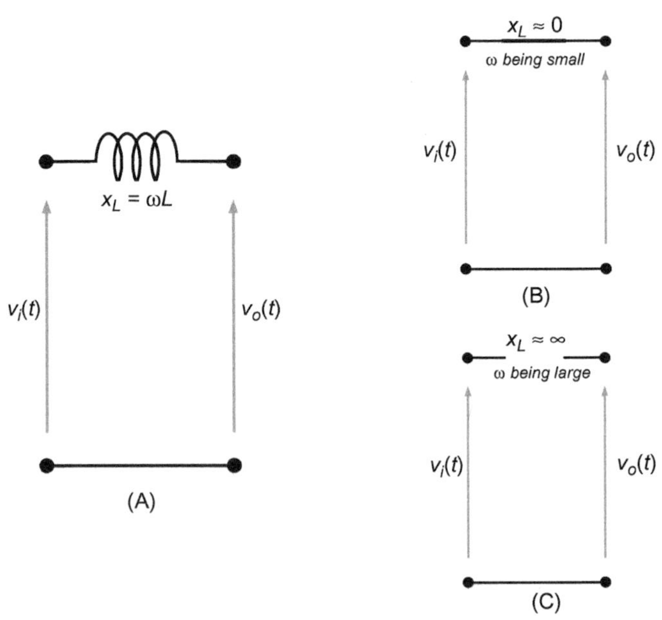

FIGURE 1.12 An inductor as a *low-pass filter*.

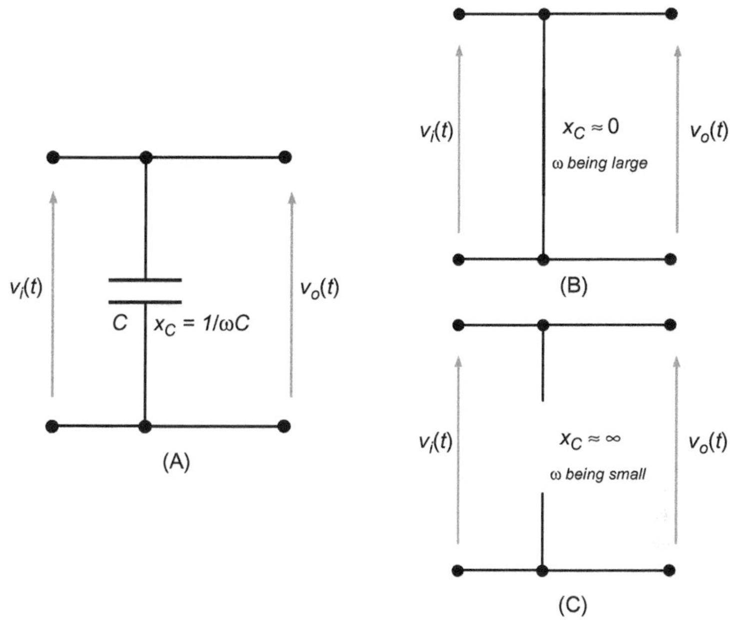

FIGURE 1.13 A capacitor as a *low-pass filter*.

assume that a certain impedance X_o is required to block a sinusoidal waveform using an inductor. Now, if we have two sinusoidal waveforms of frequencies f_1 and f_2, the respective values of inductances L_1 and L_2 required for an impedance of X_o would be:

$$X_o = 2\pi f_1 L_1 = 2\pi f_2 L_2, \tag{1.10}$$

which gives,

$$\frac{L_1}{L_2} = \frac{f_2}{f_1} \tag{1.11}$$

Eq. (1.11) indicates that the greater the frequency, the lower the value of inductance, i.e., if $f_2 > f_1$, then $L_2 < L_1$. Hence, *suppressing* a sinusoid of large frequency requires an inductor (and/or capacitor) of less value as compared to a sinusoid of low frequency. Therefore, we say that **filtering out higher-order harmonics is much easier than lower-order harmonics**. So, going back to the harmonic spectrum of the square wave as shown in Fig. 1.11, we wish that the lower-order harmonics could be "shifted" towards the higher-order side and then they can be filtered out with a less bulky filter. This is actually possible with a technique called PWM, which is the topic of discussion in Section 1.6.

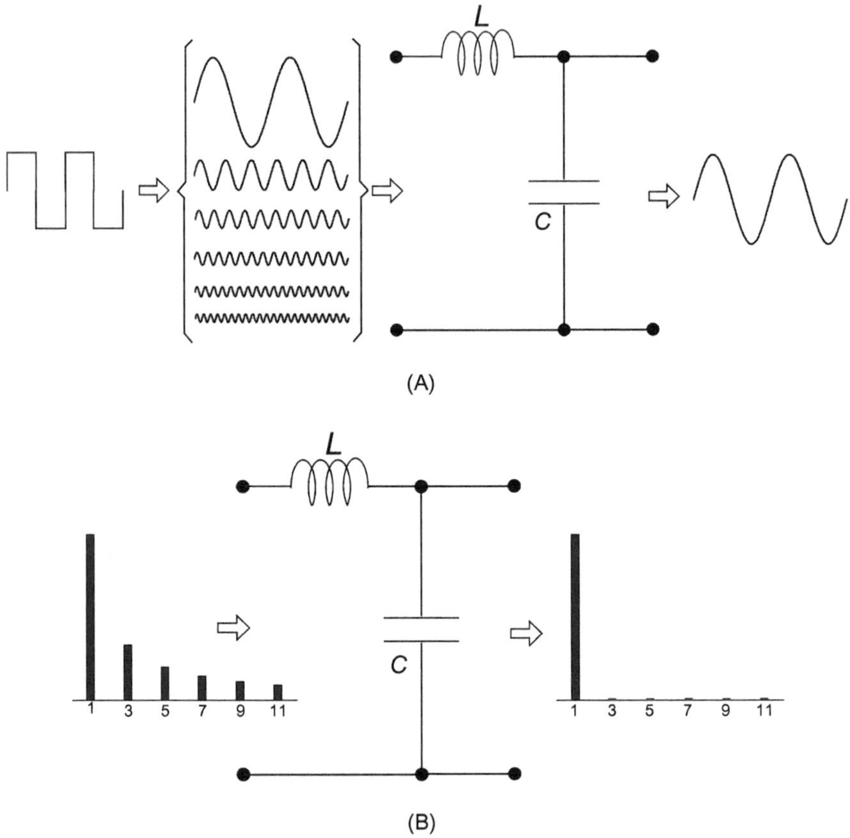

FIGURE 1.14 Concept of a low-pass filter.

1.6 SINE-TRIANGLE PWM

So far, we have discussed that the expected output waveform of an inverter is a sinusoidal one, though what we get is far from it because the waveform obtained from a discrete operation is bound to be discrete. We have also discussed that a square waveform, though alternating in nature, would require a large filter because of the dominant presence of lower-order harmonics. Sine-triangle PWM is a methodology which helps in better "imitation" of a sine waveform. In the process, as will be seen, the content of lower-order harmonics is shifted to the higher-order side and the filters become less bulky.

In Section 1.3, we saw that switches of the H-bridge topology are operated for a duration equal to that of half of the time period of the output waveform. That is to say, if a square waveform of frequency 50 Hz is to be obtained, then switches S_1 and S_4 conduct simultaneously for 10 ms and S_2 and S_3 conduct simultaneously for the next 10 ms, and such cycles are repeated. For a better imitation of a sine wave, a different switching pattern is followed in PWM, as shown in Fig. 1.15. In this method, a triangular waveform is taken as the carrier signal and a sine wave as the reference signal. Both signals are fed into a comparator. In this scheme the reference and carrier signals are continuously compared and for all instants when the value of the reference signal is greater than the carrier signal, switches S_1 and S_4 are switched ON. Otherwise, S_2 and S_3 are switched ON. As can be seen, in one cycle of the reference signal each switch is operated multiple times.

The output voltage waveform for the H-bridge inverter when operated with a sine-triangle PWM with 1 kHz triangular carrier and 50 Hz sinusoidal reference is shown in Fig. 1.16A. The harmonic profile of the PWM waveform is shown in Fig. 1.16B. An important parameter is the ratio of the amplitude of the carrier waveform to that of the reference waveform. This ratio is referred to as the **modulation index, *M***. The harmonic profile of the PWM voltage waveform now shows that the lower-order harmonics are absent. However, the THD at 145.82% is much higher as compared to the THD of the square waveform at 48.34%. Thus, though the PWM

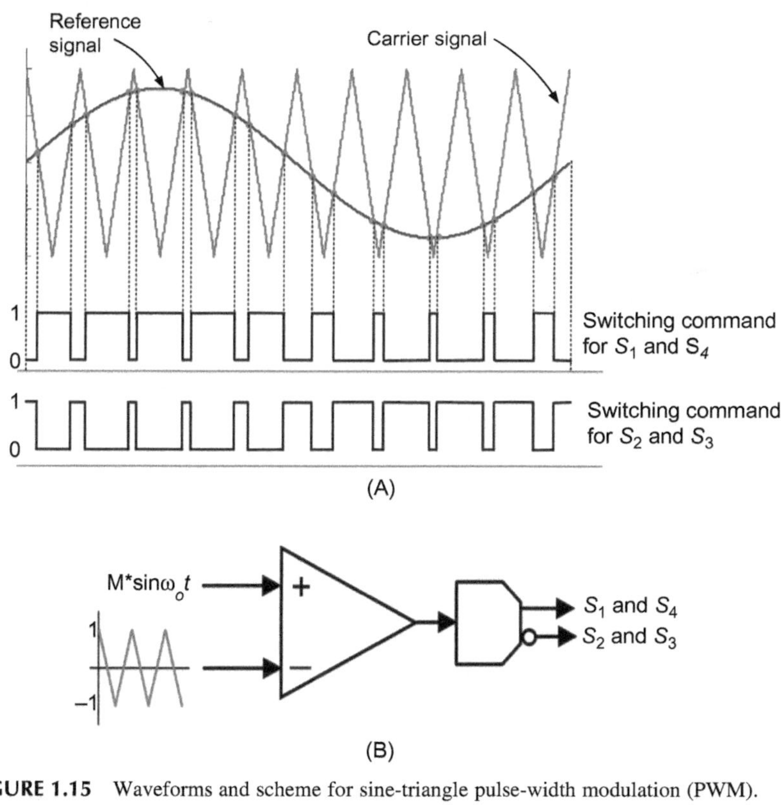

FIGURE 1.15 Waveforms and scheme for sine-triangle pulse-width modulation (PWM).

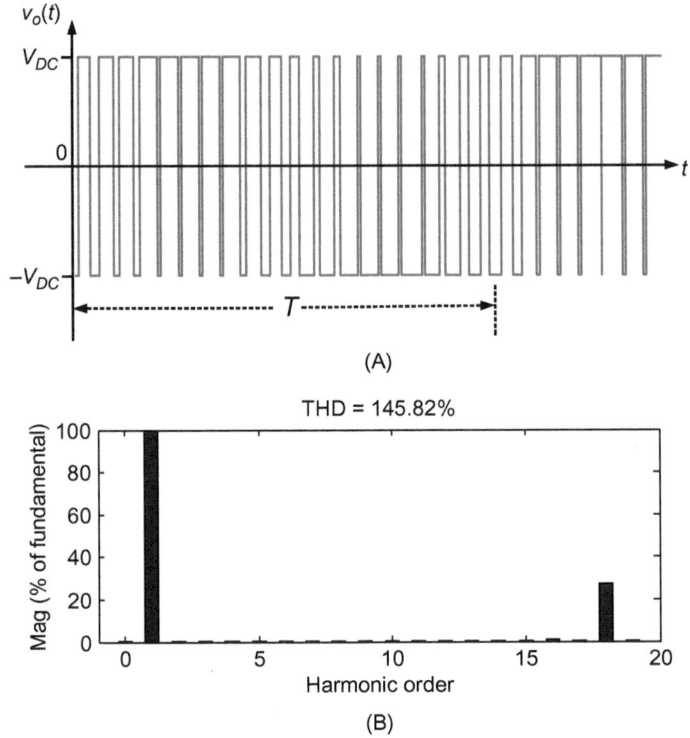

FIGURE 1.16 A two-level PWM voltage waveform and its harmonic profile.

waveform is more "distorted" as compared to the square waveform, the harmonic profile of the former is better from the point of view of filter size. Hence, the PWM waveform is the preferred one. Further discussions on PWM are given in later chapters.

1.7 POWER SWITCH REQUIREMENTS

When the H-bridge inverter was discussed in Section 1.3, switches S_1, S_2, S_3, and S_4 were only shown symbolically. In this section, the power device configuration of these switches is discussed. A "switch" is meant to be a controlled device, which can be turned ON and OFF "at will." Such devices are "fully controlled" power switches which can be controlled with the "gate" terminal. Practically, insulated-gate bipolar transistors (IGBTs) and metal-oxide semiconductor field effect transistors (MOSFETs) are the most commonly used power switch devices. When turned ON, these devices have the capability to conduct current in only one direction. The AC load can be resistive, inductive, or capacitive in nature. For the H-bridge topology of Fig. 1.6, a case of an inductive load is illustrated in Fig. 1.17A in which the

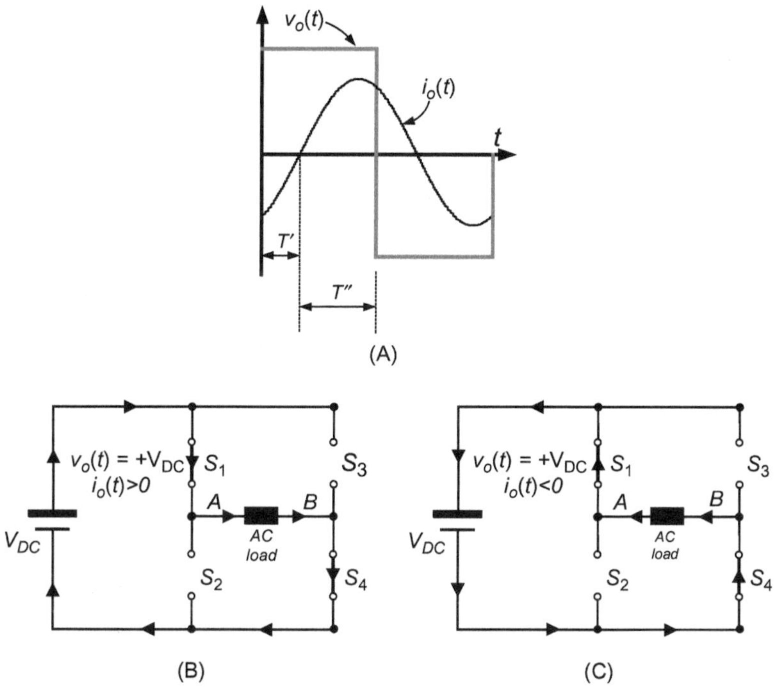

FIGURE 1.17 (A) Voltage and current waveforms for an inductive load; (B) equivalent circuit of H-bridge inverter for duration T''; (C) equivalent circuit of H-bridge inverter for duration T'.

load current lags behind the load voltage. Time durations T' and T'' are identified in which the load current is negative and positive, respectively, while the voltage is positive. Based on the discussion in Section 1.3, we know that a positive load voltage means that switches S_1 and S_4 need to remain ON. For the duration T', both switches need to conduct current in one direction and, for duration T'', the direction of current is the exact opposite, as illustrated in Fig. 1.17B and C. Hence, the switches should be **bidirectional conducting** in nature. Moreover, as seen in Fig. 1.7, whenever these switches are OFF, the polarity of voltage across them remains the same. Therefore, these switches should be of a **unidirectional blocking** nature. Thus, we conclude that the overall characteristic of power switches for an H-bridge inverter should be **bidirectional-conducting-unidirectional-blocking (BCUB)**. Such a power switch is illustrated in Fig. 1.18A in which there is a fully controlled "transistor" device (here an IGBT) and a diode connected to it in an antiparallel direction. Of course, depending on the topology there may be a requirement for a switch with the capability to block voltages of both polarities. Such power switches are called **bidirectional-conducting-bidirectional-blocking (BCBB)** switches. Topologies requiring such

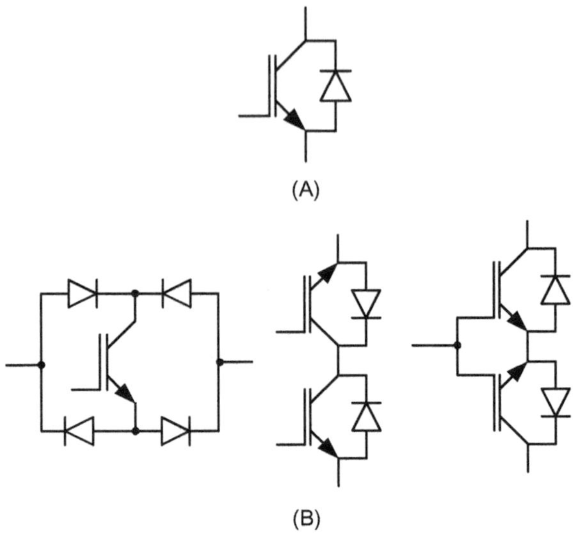

FIGURE 1.18 (A) A bidirectional-conducting-unidirectional-blocking (BCUB) power switch; (B) some configurations for bidirectional-conducting-bidirectional-blocking (BCBB) power switches.

switches are discussed in later chapters. However, for the sake of completion of this discussion, some BCBB configurations are illustrated in Fig. 1.18B.

1.8 SUMMARY

In this chapter, fundamental concepts which will help in a better understanding of multilevel inverters are discussed. Power electronics is defined in terms of the way semiconductor devices are used (i.e., as switches and not in their linear region of operation). A basic H-bridge topology is also discussed along with the power switch requirements. In addition, waveform analysis and concepts of PWM are introduced as "needs" for better performance of an inverter in terms of filter requirements. Technical terms which are widely used in literature pertaining to power electronics and multilevel inverters are also discussed in this chapter.

REFERENCES

[1] B.K. Bose, Global energy scenario and impact of power electronics in 21st century, IEEE Trans. Ind. Electron. 60 (7) (2013) 2638–2651.

[2] J.D. van Wyk, F.C. Lee, On a future for power electronics, IEEE J. Emerg. Sel. Top. Power Electron 1 (2) (2013) 59–72.

[3] M.A.J. Sathik, S.H.E.A. Aleem, R. Kannan, A.F. Zobaa, A new switched DC-link capacitor-based multi-level converter (SDC^{2}MLC), Electr. Power Compon. Syst. 45 (9) (2017) 1001–1015.

Chapter 2

Basics of Multilevel Inverters

2.1 INTRODUCTION

Chapter 1, Basics of Inverters, discussed H-bridge topology. Power switch requirement was also discussed and it was shown that for implementation of an H-bridge inverter, S_1, S_2, S_3, and S_4 are required to be of a bidirectional-conducting-unidirectional-blocking nature. It was also argued that a switch with a transistor part (such as an insulated-gate bipolar transistor) and a diode in antiparallel is suitable for the purpose, as shown in Fig. 2.1.

We also discussed obtaining voltages $+V_{DC}$ and $-V_{DC}$ across the load using two *states*. These states are again shown in Fig. 2.2A and 2.2B. Before discussing the concept of a multilevel inverter (MLI), the following points are highlighted:

1. We saw in Chapter 1, Basics of Inverters, that for an H-bridge inverter, the blocking voltage (or the minimum voltage rating of power switches) is equal to the operating voltage (or the maximum value of voltage at the load terminals).
2. We also saw that for a load fed with an H-bridge inverter, the *dv/dt* stress per unit was calculated to be:

$$\frac{dv}{dt}\text{stress in p.u. for a two}-\text{level waveform}=\lim_{\Delta t\to 0}\frac{2}{\Delta t} \tag{2.1}$$

Now, for H-bridge topology, two more states are possible, as shown in Fig. 2.2C and 22.D. In *State 3*, switches S_1 and S_3 are simultaneously turned ON such that voltage obtained across the load is 0 V. In *State 4*, switches S_2 and S_4 are simultaneously turned ON to again obtain a load voltage of 0 V.

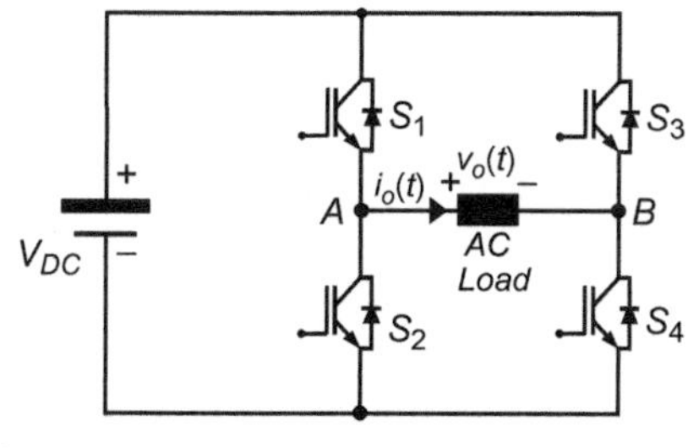

FIGURE 2.1 H-bridge inverter.

Multilevel Inverters. DOI: https://doi.org/10.1016/B978-0-12-812448-2.00002-7

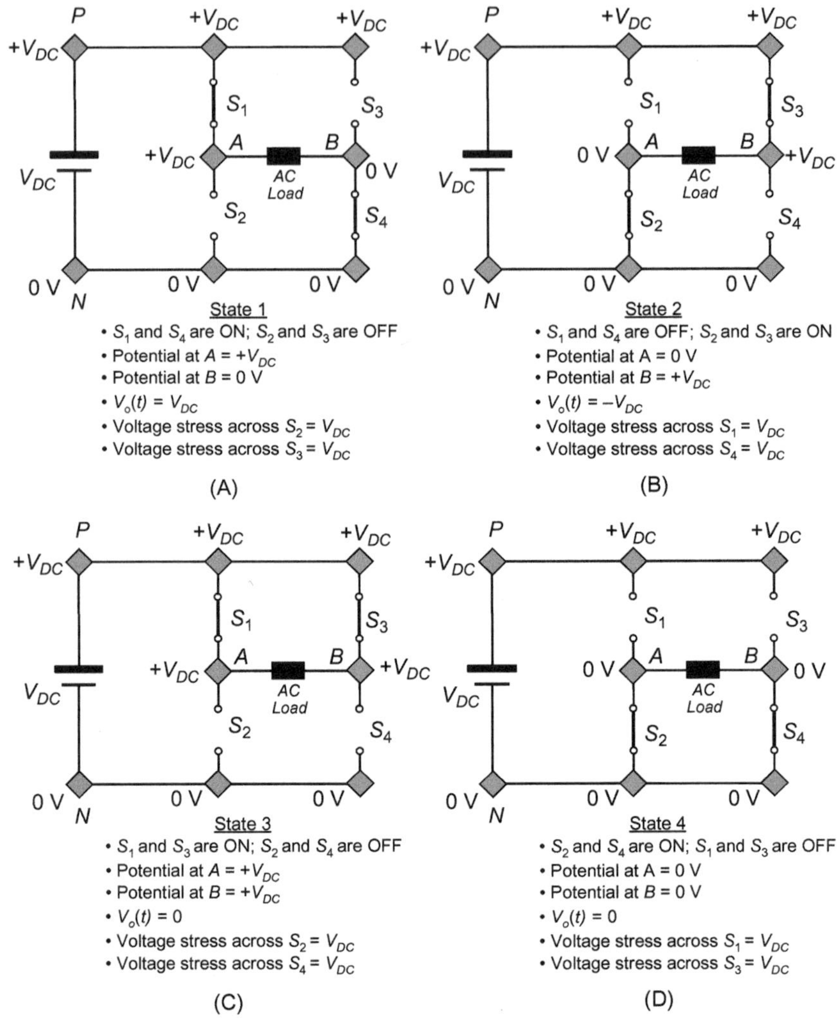

FIGURE 2.2 Four valid states for an H-bridge inverter.

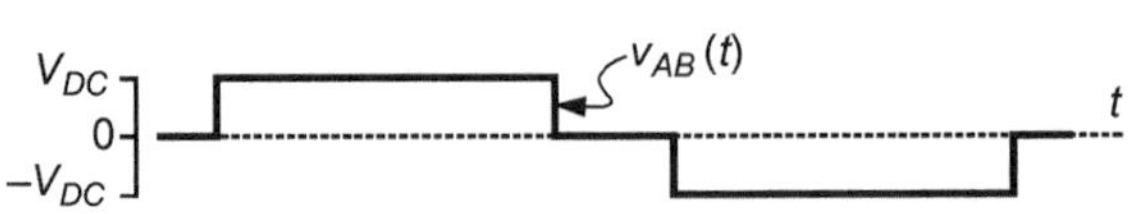

FIGURE 2.3 A three-level waveform obtained from an H-bridge inverter.

States 3 and *4* give the same output voltage, though they are distinct in terms of switching combinations. Such states are called **redundant states**. If we also include zero level in the operation of an H-bridge topology, a waveform as shown in Fig. 2.3 can be obtained.

This waveform has three distinct levels: $+V_{DC}$, 0, and $-V_{DC}$ and hence it is called a **three-level waveform**. One very important advantage of this waveform is that the *dv/dt* stress across the load is reduced by half as compared to the two-level waveform. That is to say,

$$\frac{dv}{dt}\text{stress in p.u. for a three}-\text{level waveform} = \lim_{\Delta t \to 0} \frac{1}{\Delta t} \qquad (2.2)$$

A waveform which has three or more levels is called a **multilevel waveform**, and a topology which is capable of synthesizing a multilevel waveform is called an MLI. The initial advantage of a multilevel waveform is reduced *dv/dt* stress on the load. The greater the number of levels in a waveform, the lower will be the *dv/dt* stress on the load (or the filter). In Section 2.2, the concept of MLIs is introduced and their advantages are demonstrated. Later in the chapter, conventional topologies are described and their limitations discussed.

2.2 MULTILEVEL INVERTERS

Let us consider three separate H-bridge inverters, with each having input DC voltage equal to $V_{DC.}$ As illustrated in Fig. 2.4, the load terminal voltages of these bridges are $v_{o,1}(t)$, $v_{o,2}(t)$, and $v_{o,3}(t)$. Each of these bridges can be operated as a three-level inverter. Note that the blocking voltage for the power switches of each bridge is V_{DC}. These bridges can be operated in such a manner that the widths of positive half cycles and negative half cycles are "calculatedly" different, as shown in Fig. 2.4. Now, if we connect the three bridges in series, as shown in Fig. 2.5A, the load voltage $v_o(t)$ will be the sum of $v_{o,1}(t)$, $v_{o,2}(t)$, and $v_{o,3}(t)$ and as can be seen in Fig. 2.5B, the load voltage is "stepped" in nature, consisting of seven levels ($\pm V_{DC}$, $\pm 2V_{DC}$, $\pm 3V_{DC}$, and 0). This seven-level waveform seems to be a better imitation of a sine wave as compared to a square waveform or even a three-level waveform. Very importantly, in this case, the blocking voltage (equal to V_{DC}) is one-third of the operating voltage (equal to $3V_{DC}$) and hence power switches of lesser voltage rating can be used to synthesize a much higher voltage waveform. Compare this with the case of a single H-bridge discussed in Chapter 1, in which the blocking voltage was equal to the operating voltage.

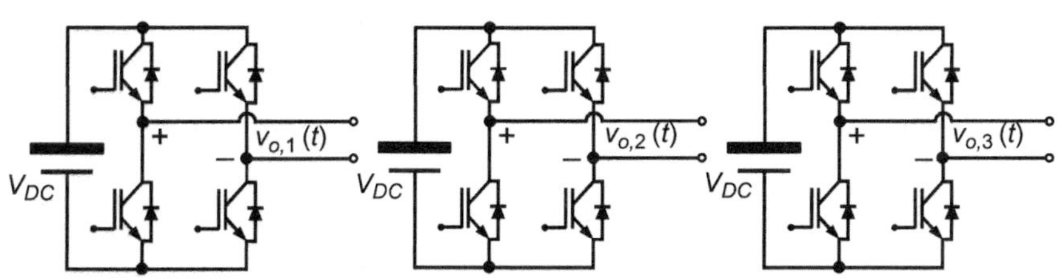

FIGURE 2.4 Three H-bridges with an input voltage of V_{DC} each.

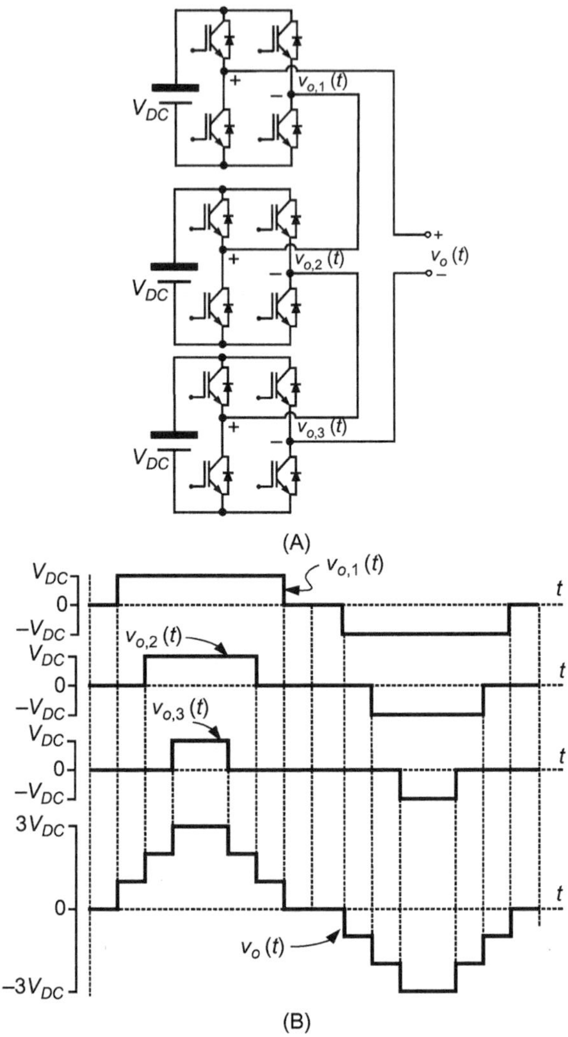

FIGURE 2.5 (A) Three H-bridges, with three-level output, connected in series; (B) individual output waveforms of the H-bridges and the load voltage waveform.

That is to say, for an operating voltage of 600 V, the full bridge inverter will require power switches rated at 600 V, whereas an H-bridge-based seven-level inverter will require power switches rated at 200 V.

In addition, the *dv/dt* stress on the load is ,

$$\frac{dv}{dt}\text{stress in p.u. for a seven} - \text{level waveform} = \lim_{\Delta t \to 0} \frac{1}{3\Delta t} \tag{2.3}$$

Thus, *dv/dt* stress on the load for a seven-level waveform is one-sixth of that of a two-level waveform.

There are many more advantages to MLIs in addition to those which have been demonstrated so far [1]. The advantages of MLIs can be categorized as:

1. Advantages due to multilevel waveform;
2. Advantages due to multilevel topology.

One might argue that these two advantages are complementary. After all, a multilevel topology *will* synthesize a multilevel waveform. However, we are of the view that as new topologies are emerging, they should be analyzed critically. A topology may be able to synthesize a multilevel waveform but may *not* offer *all* the advantages of an MLI. Clear discrimination is possible when the benefits are clearly understood in terms of the waveform as well as the topology.

2.2.1 Advantages of MLIs on Account of the Waveform

By virtue of the stepped waveform, the following advantages are associated with MLIs:

1. The staircase waveform exhibits better quality and a better harmonic profile. Thus, the filter requirements can be drastically reduced (or even eliminated). In Fig. 2.6, three-level, five-level, and seven-level waveforms

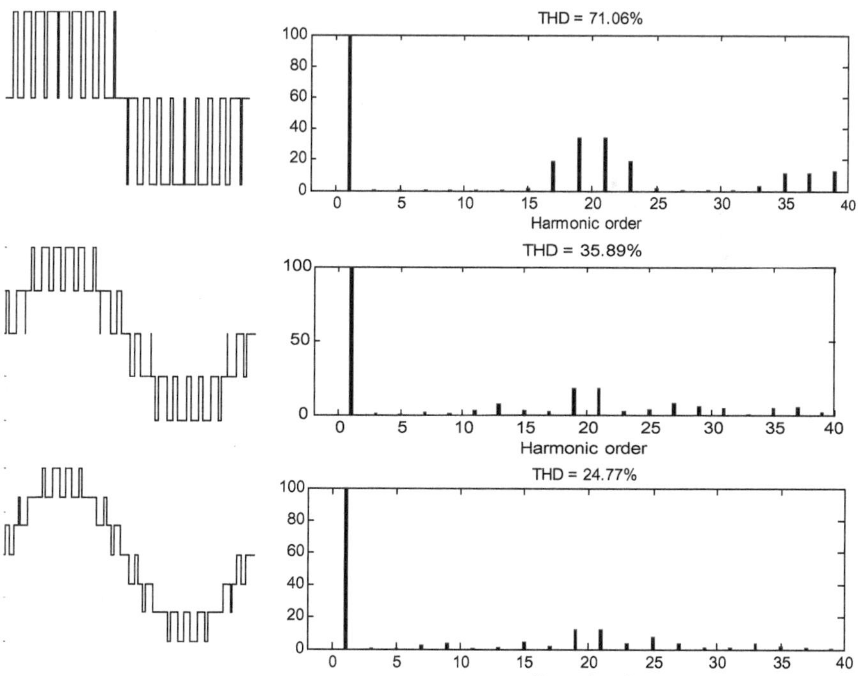

FIGURE 2.6 Three-level, five-level, and seven-level waveforms with their respective harmonic profiles. *THD*, total harmonic distortion.

are shown with their respective harmonic profiles. Each waveform is obtained with carrier waveforms of 1 kHz and a modulation index of 0.85. It can be observed that as the number of levels increases, not only does the total harmonic distortion reduce, but lower-order harmonics are also reduced in the harmonic profile.

2. The *dv/dt* stress on the load is greatly reduced. As demonstrated above, as the number of steps or levels increases, the *dv/dt* stress on the load is reduced, thereby positively impacting on the lifetime of the load. Electromagnetic compatibility problems are also reduced.

3. The staircase waveform also offers more degrees of freedom to control the output waveform, in terms of number of levels, step-size, and switching angles, apart from the amplitude and frequency modulation indices and phase.

4. MLIs produce much a lower common mode voltage (CMV) and thus the stress in the bearings of a motor connected to a multilevel motor drive can be reduced.

5. MLIs can be modulated with both fundamental frequency and high-frequency switching schemes. This point will be further explored in later chapters.

2.2.2 Advantages of MLIs on Account of Topology

In terms of topology for multilevel DC to AC conversion, following are the typical advantages:

1. The voltage stresses on the semiconductor devices are much less than the overall operating voltage, thus enabling the usage of traditional semiconductor devices. Also, more importantly, series-parallel connection of power switches is not required.

2. Many topologies of MLIs allow multiple redundant states for a number of output voltage levels. Such phase and/or line redundancies available in multilevel topologies can be utilized to program a fault-tolerant operation. These redundant states can also be used to equalize the power contributed by the input DC sources to enhance their lifetime and reduce maintenance. In addition, power switches can be operated at optimal switching frequencies using redundant states. Redundant states in three-phase topologies are also used to minimize the CMV. These points are explained in detail in later chapters along with examples of applications.

3. MLIs can draw input current with low distortion.

4. Renewable energy sources, such as photovoltaic, wind, and fuel cells, can be easily interfaced to a multilevel converter system and can be controlled for equal load sharing amongst the input sources. This is because an MLI requires multiple input DC levels to synthesize a stepped waveform.

These advantages are summarized in Fig. 2.7.

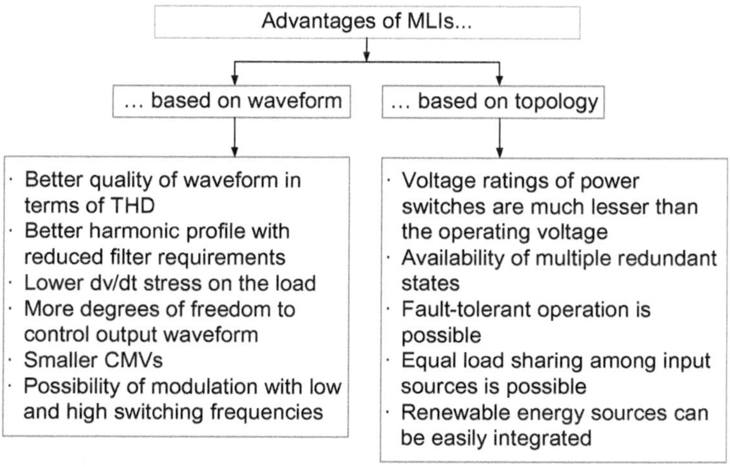

FIGURE 2.7 Advantages of MLIs. *THD*, total harmonic distortion.

2.3 CONVENTIONAL TOPOLOGIES

We discussed in Section 2.2 that by connecting multiple H-bridges in series and modulating them appropriately, a multilevel waveform is synthesized. There are many such "circuit configurations or topologies" capable of synthesizing a multilevel waveform. Many such topologies are discussed in this book. Three topologies, however, have acquired the status of being "classic or conventional" topologies because an expansive literature is available describing them in detail. These topologies are: cascaded H-bridge (CHB), neutral point clamped (NPC), or diode-clamped and flying capacitor (FC) MLIs. These are discussed in Sections 2.3.1–2.3.3.

2.3.1 CHB Inverter and Modulation Strategies

The topology discussed in Section 2.2 to introduce the concept of the synthesis of a multilevel in which multiple H-bridges are connected in series is the CHB inverter. In the mid-1970s, the first patent describing this converter topology capable of producing multilevel voltage from various DC voltage sources was published by Baker and Bannister [2]. The topology comprises single-phase inverters connected in series as depicted in Fig. 2.5A and it is known as a series-connected H-bridge inverter or CHB inverter. Each H-bridge is generally called a "cell." Thus, each cell has its own input DC level and the input DC levels of various cells are electrically isolated from one another. In order to understand the modulation of the CHB topology, we first understand how the pulse-width modulation (PWM) of a single-cell (or three-level) H-bridge inverter is carried out.

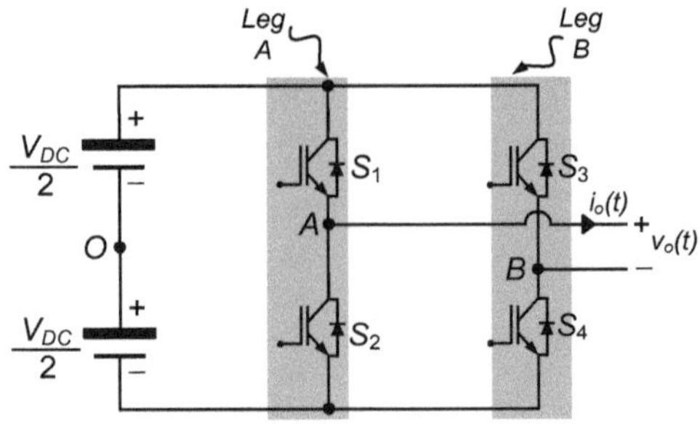

FIGURE 2.8 One "cell" of a CHB-MLI.

The H-bridge, as shown in Fig. 2.8, has two "legs." For the sake of explanation, the DC link voltage is separated into two parts and a common point "O" is indicated. Leg-A consists of two power switches, S_1 and S_2. When S_1 is turned ON, Point A gets connected to the positive DC rail and $v_{AO}(t) = V_{DC}/2$. When S_2 is turned ON, point A gets connected to the negative DC rail and $v_{AO}(t) = -V_{DC}/2$. Also, S_1 and S_2 cannot be turned ON simultaneously as this will directly short-circuit the DC source. Similarly, leg-B has complementary switches S_3 and S_4 and switching them ON gives $v_{BO}(t) = V_{DC}/2$ and $v_{BO}(t) = -V_{DC}/2$, respectively. These two legs are modulated with a sine-triangle PWM scheme as shown in Fig. 2.9.

Observe from Fig. 2.9A that a common triangular carrier signal is used for both legs, while two out-of-phase sinusoidal references are used. Carrier and reference waveforms are shown in Fig. 2.9B and 2.9C. Once both legs are accordingly operated, the load voltage $v_{ab}(t)$ is obtained as the difference between $v_{AO}(t)$ and $v_{BO}(t)$. The three waveforms are shown in Fig. 2.10.

If the CHB topology is composed of two "cells" as shown in Fig. 2.11, then two carrier signals are used, one for each cell. These carrier signals are "phase-shifted" at an electrical angle of 90 degrees. The scheme and waveforms are shown in Fig. 2.12.

Both cells give a three-level output and cascading gives a five-level PWM waveform across the load, as depicted in Fig. 2.13.

If there are "n" cells in a CHB inverter, then the "phase-shifted carrier scheme" can be extended by using carrier signals with a phase shifting at a $180°/n$ electrical angle. Of course, for a three-phase inverter based on CHB topology, reference signals also need to be phase-shifted by a 120° electrical angle.

2.3.2 Diode-Clamped Structure and Modulation Strategies

NPC topology, which is also known as diode-clamped topology, started receiving attention when, in the 1980s, Nabae et al. demonstrated the implementation of an NPC inverter along with experimentation, using a

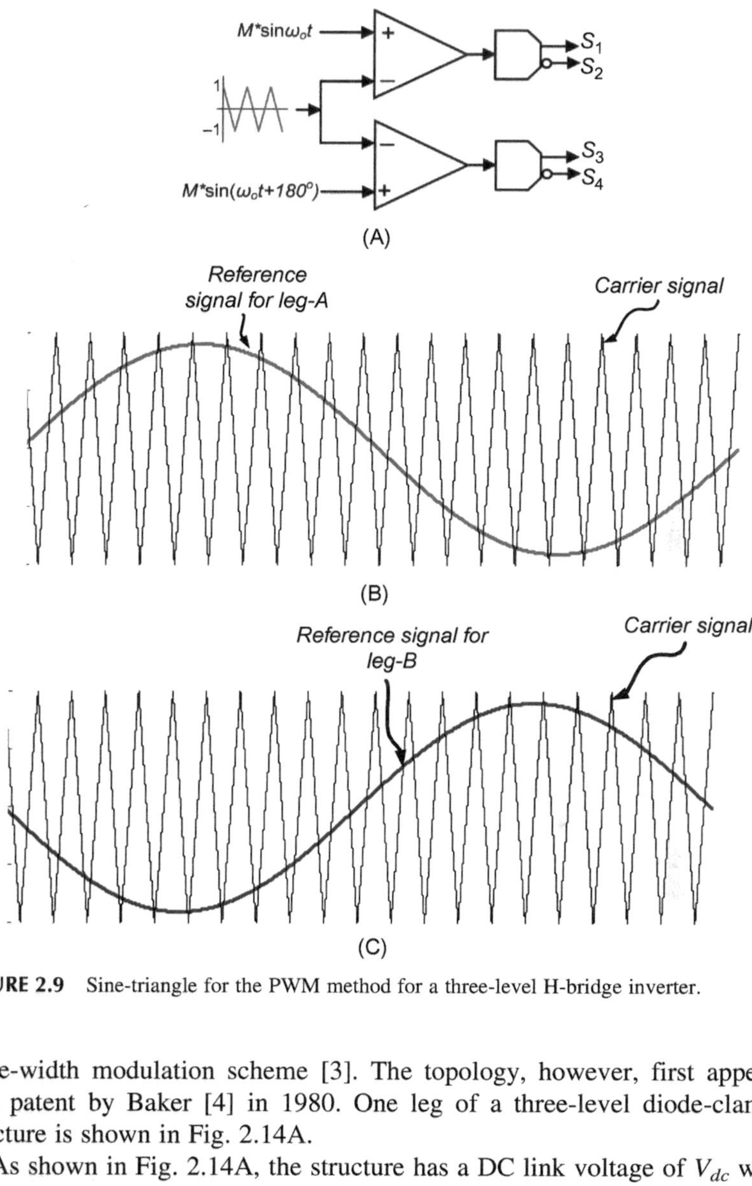

FIGURE 2.9 Sine-triangle for the PWM method for a three-level H-bridge inverter.

pulse-width modulation scheme [3]. The topology, however, first appeared in a patent by Baker [4] in 1980. One leg of a three-level diode-clamped structure is shown in Fig. 2.14A.

As shown in Fig. 2.14A, the structure has a DC link voltage of V_{dc} which is generally halved using capacitors, but here it is shown with two DC sources. The leg consists of four bidirectional-conducting-unidirectional-blocking power switches. The clamping to the common point "o" is done with the help of two diodes, D_1 and D_2. Positive and negative DC rails are shown with "p" and "n," respectively. In order to understand the workings, we replace the power switches with symbolic switches as shown in Fig. 2.14B. There are three valid states for this structure, as depicted in

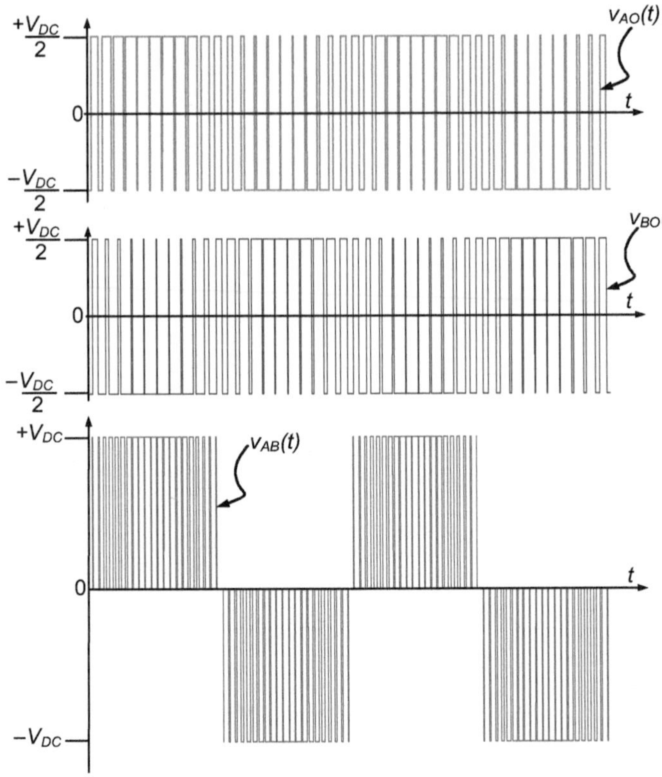

FIGURE 2.10 Waveforms for a three-level H-bridge inverter.

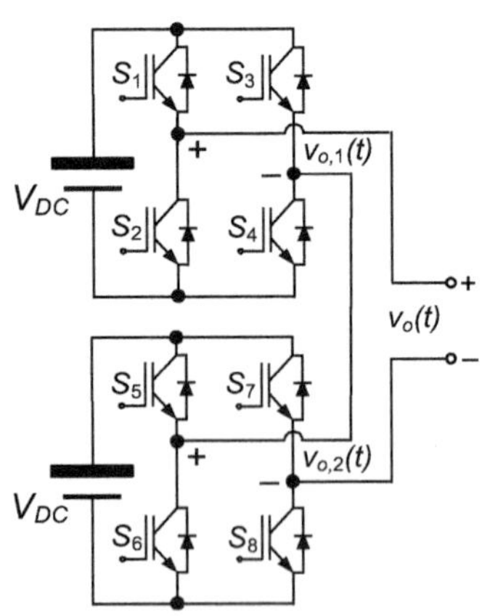

FIGURE 2.11 A CHB inverter with two "cells."

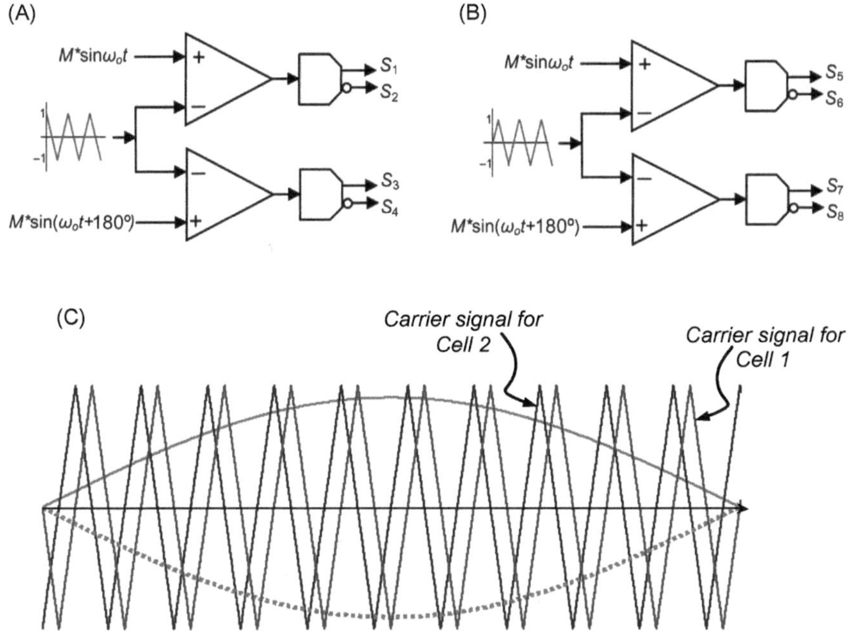

FIGURE 2.12 Modulation scheme for CHB with two cells.

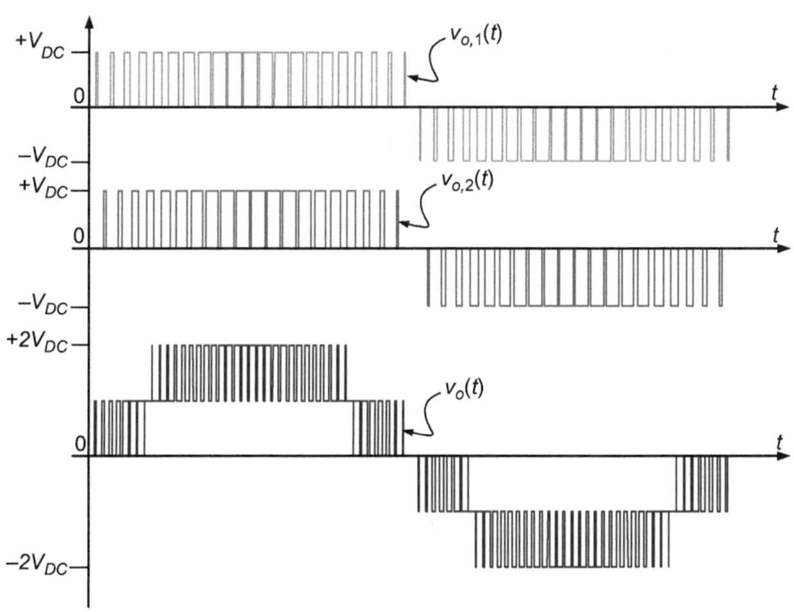

FIGURE 2.13 Voltage waveforms for a CHB inverter with two cells.

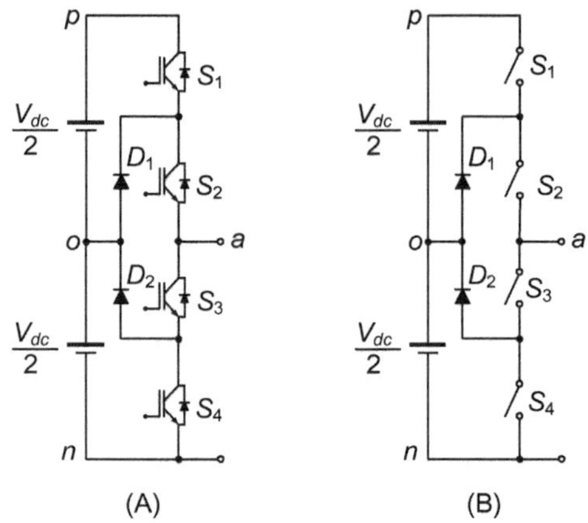

FIGURE 2.14 (A) One leg of a three-level diode-clamped inverter; (B) leg shown with "switches."

Fig. 2.15. For analysis, point "o" is chosen as the reference node with a potential of 0 V. Accordingly, the potentials at p and n are $+V_{dc}/2$ and $-V_{dc}/2$. The potential difference between Point "a" and "o" in the circuit is considered to be the load voltage, $v_{ao}(t)$. In State 1, S_1 and S_2 are simultaneously switched ON to obtain $v_{ao}(t) = V_{dc}/2$, as shown in Fig. 2.15A. It can be observed that the voltage stress across the devices in the OFF state is $V_{dc}/2$ which is half the DC link voltage. In State 2, two equivalent circuits are shown with directions of currents to emphasize the role of clamping diodes in the topology. When Switches S_2 and S_3 are simultaneously switched ON, then the load voltage is zero. At this time, if the load current is positive, then the current completes its path through S_2 and D_1 as depicted in Fig. 2.15B. If the load current happens to be negative, then it completes its path through S_3 and D_2, again depicted in Fig. 2.15B. Once again, the devices in the OFF state block a voltage equal to $V_{dc}/2$. In State 3, when Switches S_3 and S_4 are simultaneously turned ON, then the obtained load voltage is $-V_{dc}/2$, as shown in Fig. 2.15C. In this state also, the voltage stress across the OFF devices is $V_{dc}/2$. Hence, three levels are obtained using these three states: $\pm V_{dc}/2$ and 0.

In order to modulate this leg, the PWM scheme is as shown in Fig. 2.16A, along with the carrier and reference signals in Fig. 2.16B. A single reference is used, while two carrier signals are used. As opposed to the "phase-shifted" carrier signals used for the CHB topology, here the carrier signals are "level-shifted." That is to say, one carrier signal is above the zero reference and the other is below the zero reference. The reference signal is continuously compared with the carrier signals. Whenever the reference signal is greater than the carrier signal 1, Switch S_1 is turned ON, otherwise S_3 is turned ON. Similarly, whenever the reference signal is greater than carrier signal 2,

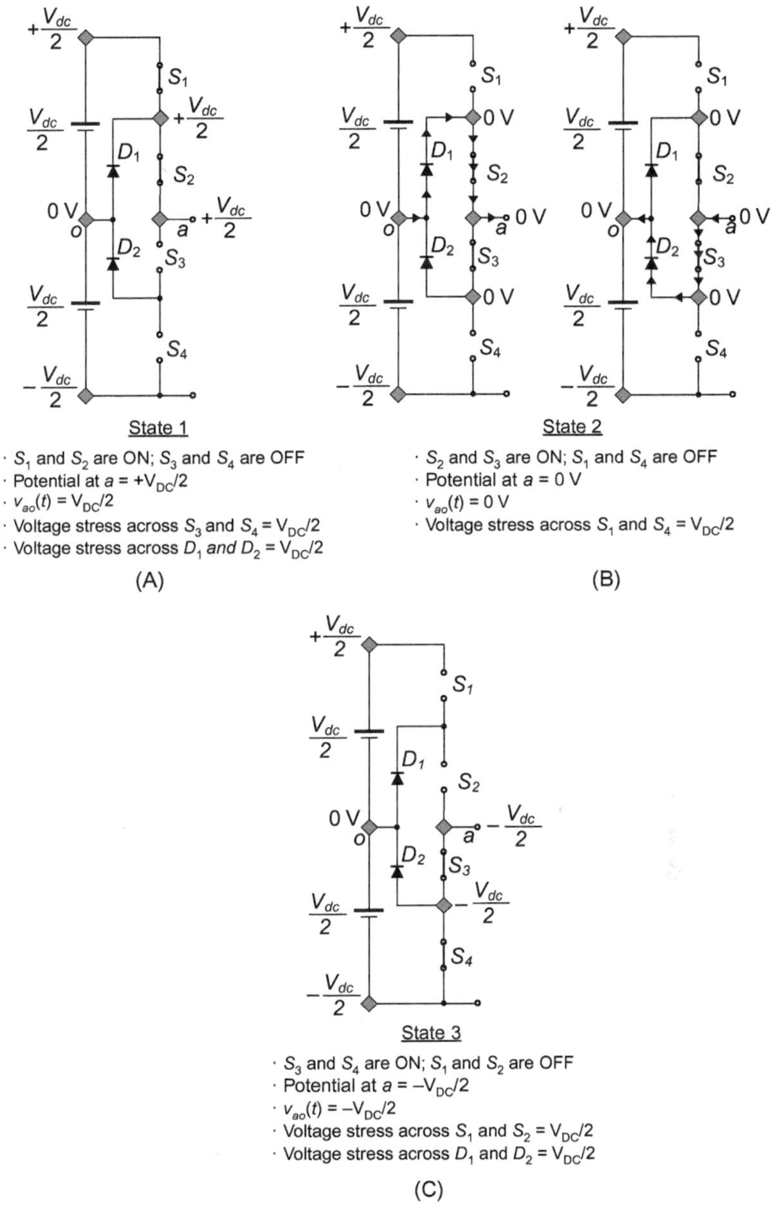

FIGURE 2.15 Working states for one leg of a three-level diode-clamped inverter.

switch S_2 is turned ON, otherwise Switch S_4 is turned ON. When operated in this manner, the resulting load voltage $v_{AO}(t)$ is as shown in Fig. 2.16C.

We have previously discussed that, for an MLI, generally, the voltage stress across the power switches is less than the operating voltage. This

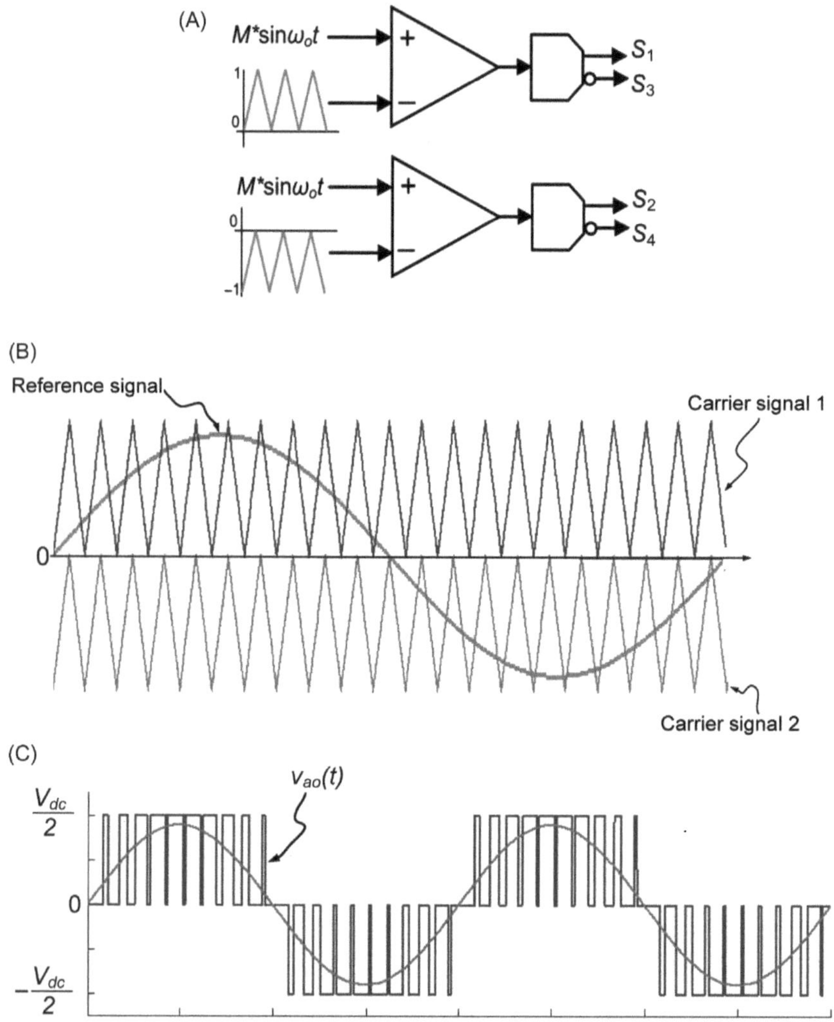

FIGURE 2.16 (A) Modulation strategy for one leg of a three-level diode-clamped inverter; (B) reference and carrier signals; and (C) output voltage waveform.

advantage can be seen here if three-phase voltage is obtained. A three-phase structure is obtained by "adding" to more three-level legs as shown in Fig. 2.17.

The three-phase structure can be modulated by modulating all three legs with level shifted carriers and sinusoidal references with a phase difference of 120°, as shown in Fig. 2.18. The corresponding carrier and reference signals are shown in Fig. 2.19. When modulated in this manner, the phase voltages have three levels, with $V_{dc}/2$ as the peak value, and they are shifted

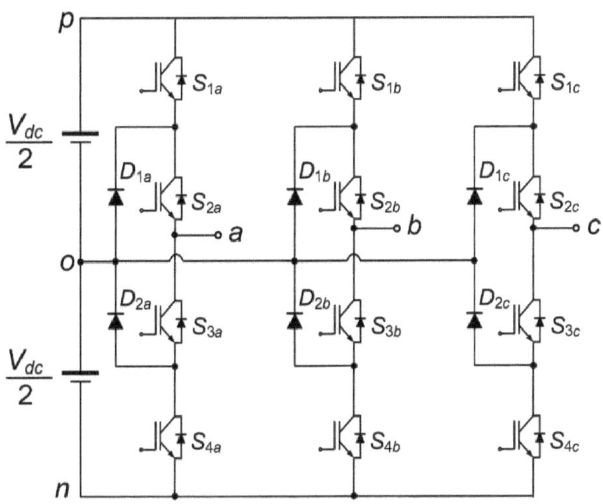

FIGURE 2.17 A three-phase diode-clamped MLI.

from one another by an angle of 120 degrees, as shown in Fig. 2.20. The line voltages shown in Fig. 2.21 are all observed to consist of five levels. These are all phase-shifted from one another at an angle of 120 degrees and have a peak value of V_{dc}, which is twice the blocking voltage requirement. Here we emphasize the relationship between the number of levels in phase and line voltages:

$$\text{Number of levels in line voltage} = 2 \times (\text{number of levels in phase voltage}) - 1 \tag{2.4}$$

In order to extend the number of levels, the structure can be systematically extended. One such example is shown in Fig. 2.22 with a single leg. The reader is encouraged to determine working valid states for this structure and to see how the clamping diodes shown have unequal blocking voltage requirements. Similarly, for modulation of diode-clamped structures with a higher number of levels, the number of carrier signals can be increased with appropriate "level-shifting" as depicted in Fig. 2.23. Later chapters contain more discussions on modulation techniques.

2.3.3 FC Structure and Modulation Strategies

The so-called FC was introduced in the late 1980s [5, 6]. This topology does not use any clamping diodes. Instead it uses capacitors which are not directly connected to the positive or negative DC rails; hence the name "floating" or "flying" capacitor (FC) topology. A leg of three-level FC topology is shown in Fig. 2.24A.

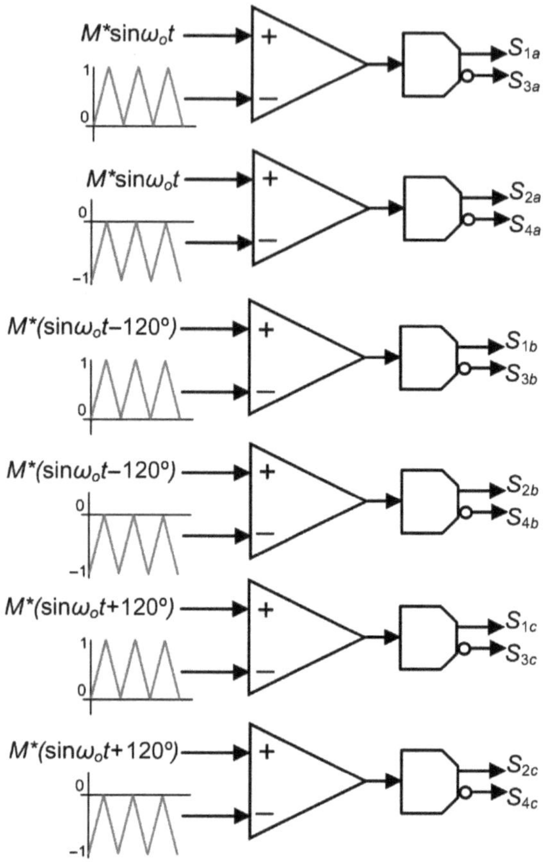

FIGURE 2.18 Modulation strategy for three-phase diode-clamped inverter with three-level legs.

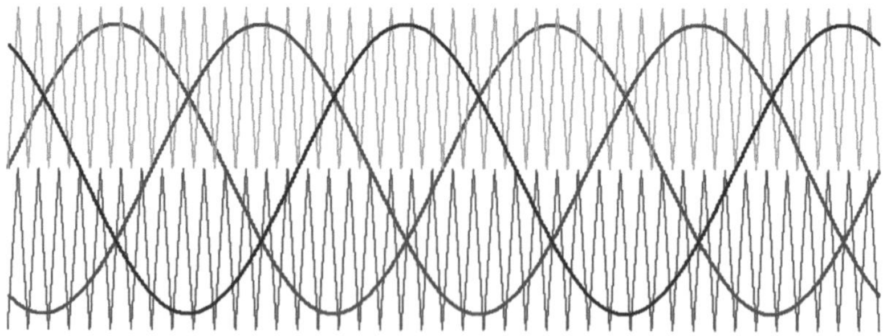

FIGURE 2.19 Triangular carrier and sinusoidal reference signals for a three-phase diode-clamped inverter with three-level legs.

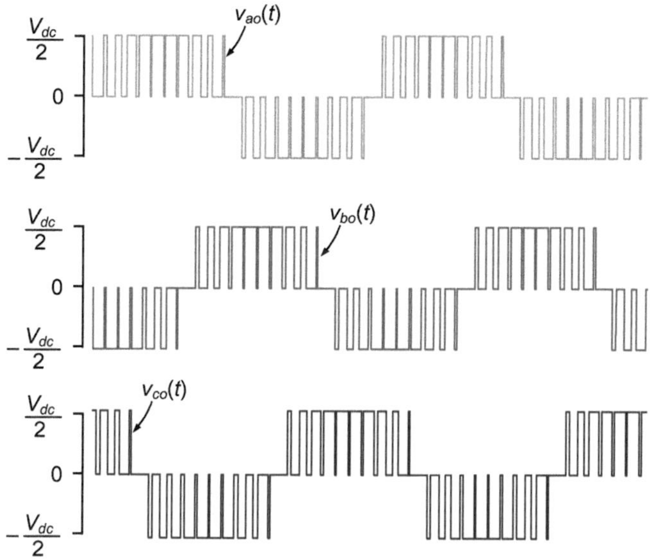

FIGURE 2.20 Phase voltage waveforms for a three-phase diode-clamped inverter with three-level legs.

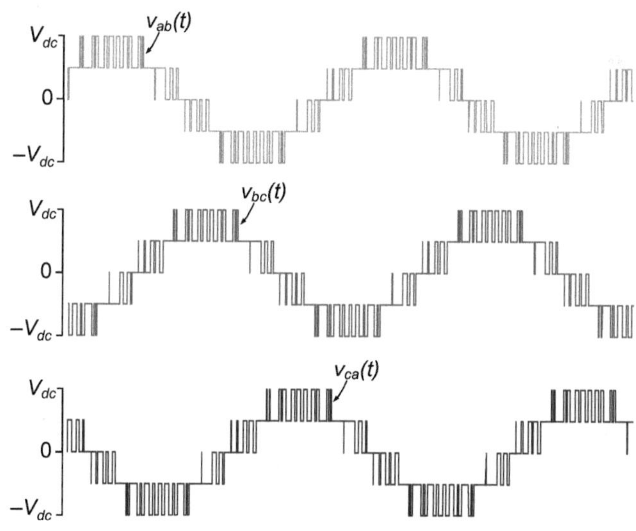

FIGURE 2.21 Line voltage waveforms for a three-phase diode-clamped inverter with three-level legs.

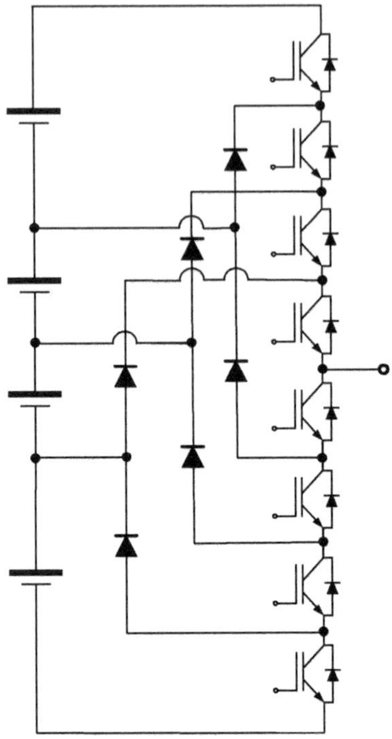

FIGURE 2.22 One leg of a five-level diode-clamped inverter.

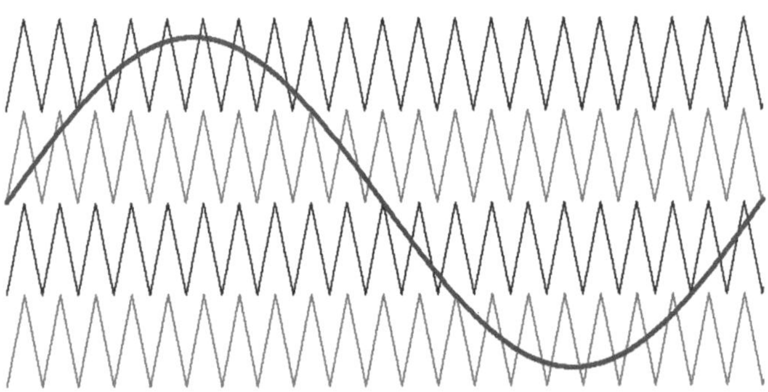

FIGURE 2.23 Multiple "level-shifted" carrier signals to obtain a five-level PWM waveform.

As shown in Fig. 2.24A, the input voltage is V_{dc}, shown with two voltage sources of value $V_{dc}/2$ each. The common point is marked as "o" and the load voltage is measured as the potential difference between points "a" and "o," $v_{ao}(t)$. The structure has four bidirectional-conducting-unidirectional-blocking switches and one floating capacitor "C." The positive and negative DC rails are marked with "p" and "n," respectively. The floating capacitor

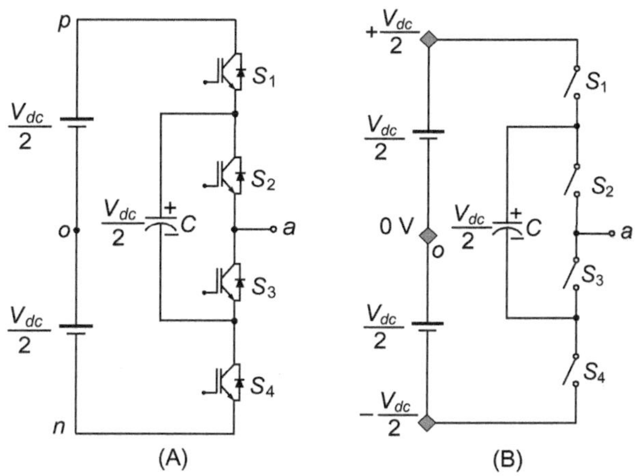

FIGURE 2.24 (A) One leg of a three-level FC MLI; and (B) equivalent representation for the purposes of analysis.

"*C*" is "precharged" to a voltage $V_{dc}/2$ with polarity as shown in Fig. 2.24A. The modulation scheme has to ensure proper maintenance of this voltage across the capacitor. For the purpose of analysis of this structure, an equivalent circuit with symbolic switches is shown in Fig. 2.24B. Point "*o*" is assumed to be the reference node with a potential of 0 V. Accordingly, the potentials at the positive and negative rails are $+V_{dc}/2$ and $-V_{dc}/2$. There are four valid states summarized in Fig. 2.25.

As can be observed from Fig. 2.25A, when S_1 and S_2 are simultaneously switched ON, a voltage of $V_{dc}/2$ is obtained across the load, while the voltage stress across the blocking switches is $V_{dc}/2$. Similarly, in State 2, shown in Fig. 2.25B, when Switches S_3 and S_4 are simultaneously turned ON, a voltage of $-V_{dc}/2$ appears across the load, while the voltage stress across the blocking devices is $V_{dc}/2$. States 3 and 4, shown in Fig. 2.25C and 2.25D, respectively, both lead to a load voltage of 0 V, although with different switching combinations, and hence they are redundant states. These redundant states are used to balance the capacitor voltages. As described earlier, the FC has to be precharged and voltages need to be balanced. Such switching procedures are discussed in later chapters, in which control is heavily based on the state selection procedure.

A three-phase inverter based on a three-level FC leg is shown in Fig. 2.26 and an extension of this topology is shown in Fig. 2.27. The reader should work out the valid states for this structure.

While a more detailed discussion on modulation schemes is given in later chapters, it should be emphasized here that, in principle, any topology can be modulated with either a level-shifted or phase-shifted carrier-based methodology, when suitably adapted.

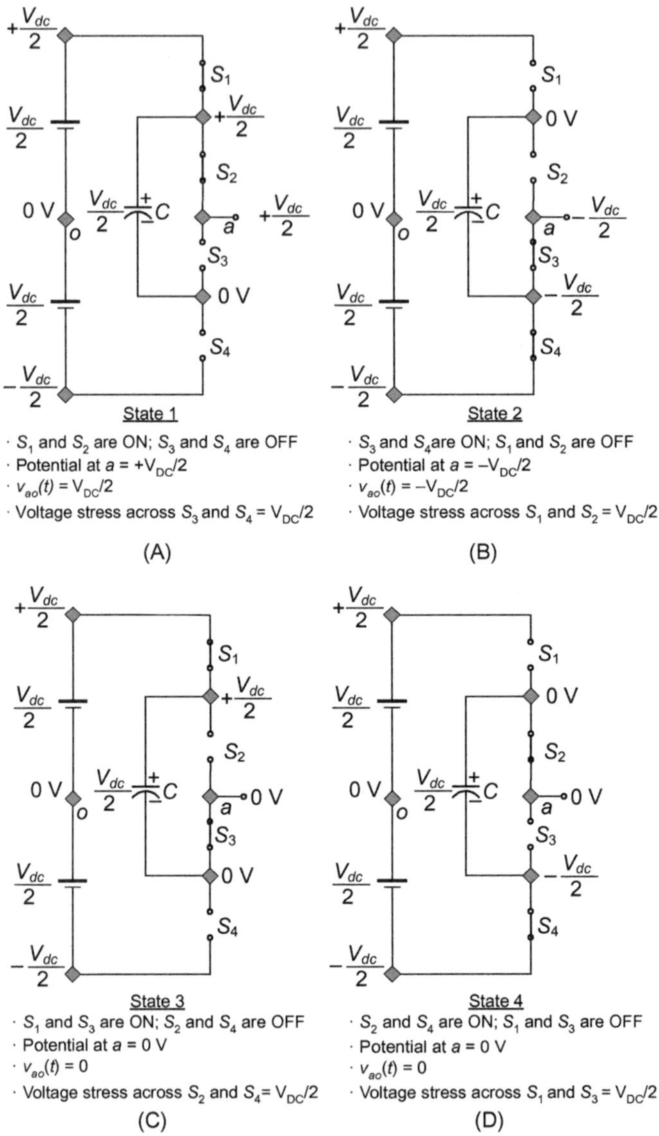

FIGURE 2.25 Valid working states for one leg of a three-level FC inverter.

2.4 ISSUES WITH CONVENTIONAL TOPOLOGIES

The fundamentals of conventional/classical topologies were discussed in Section 2.3. These topologies have been commercialized, but they have certain issues. The CHB topology requires isolated DC sources and, in the case that these sources are not readily available (e.g., in the case of AC drives),

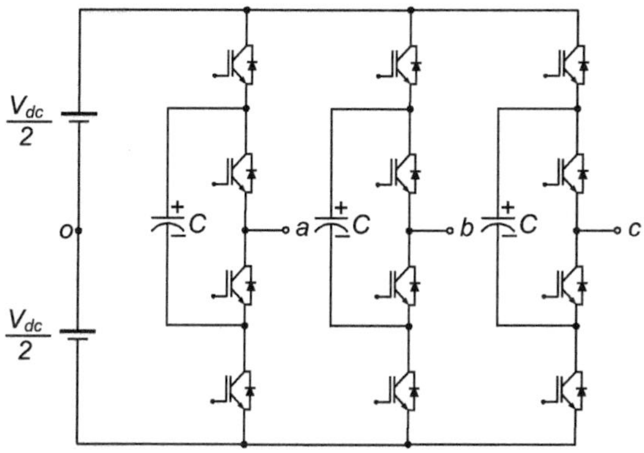

FIGURE 2.26 A three-phase FC topology with three-level legs.

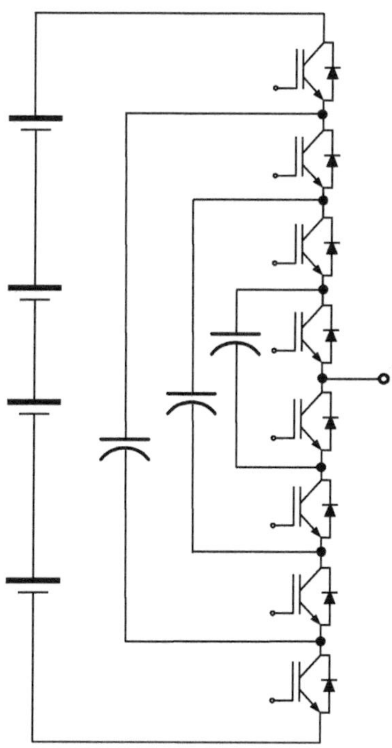

FIGURE 2.27 A five-level leg based on FC topology.

transformers with multiple secondary windings are used along with a rectifier to obtain isolated DC levels. CHB topology, of course, is highly modular in nature. For the diode-clamped inverter, a single DC link works for all three phases. But, as the number of levels increases, voltage balancing in DC link capacitors becomes difficult. Also, the switching frequencies are different for the switches of a leg, which leads to unequal lifetimes of these switches. Diode-clamped inverters, however, are highly suitable for back-to-back applications such as AC electric drives. In the case of FC topology, a single DC link is sufficient for three phases. Here too, for an increased number of levels, balancing voltages in DC link capacitors and in floating capacitors becomes difficult. In addition, the presence of a large number of capacitors puts a limit on the switching frequency.

In addition to these, there is a common issue with these topologies: for an increased number of levels, these topologies inadvertently require a large number of switches. Each power switch is accompanied by a gate driver unit and heat sink, hence the overall system becomes bulky and complex. This is *one* of the reasons behind the emergence of advanced topologies in the last few years. Detailed discussions on this will follow in later chapters.

2.5 SUMMARY

In this chapter a simplified and systematic approach is used to introduce the concept of MLIs. In addition to explanation of the concept, popular terminology related to MLIs is also discussed where appropriate. A detailed account of the three classic topologies (namely, CHB, diode-clamped, and FC topologies) is presented. With the application to these topologies, an important introduction to multiple carriers-based multilevel PWM schemes is also presented in this chapter. Level-shifted carriers-based PWM is discussed with reference to diode-clamped topology, while phase-shifted carriers-based multilevel PWM is discussed with reference to the CHB topology.

REFERENCES

[1] K.K. Gupta, A. Ranjan, P. Bhatnagar, L. Kumar, S. Jain, Multilevel inverter topologies with reduced device count: a review, IEEE Trans. Power Electron. 31 (1) (2016) 135–151.

[2] R.H. Baker, L.H. Bannister, Electric power converter, U.S. Patent 3 867 643 (1975).

[3] A. Nabae, I. Takahashi, H. Akagi, A neutral-point-clamped PWM inverter. In: Conference Record IEEE IAS Annual Meeting, Cincinnati, OH, vol. 3, pp. 761–766, September 28–October 3, 1980.

[4] R.H. Baker, Switching circuit. U.S. Patent 4210826 (1980).

[5] M. Marchesoni, M. Mazzucchelli, S. Tenconi, A non conventional power converter for plasma stabilization. In: Proceedings of the IEEE 19th Power Electronics Specialist Conference, pp. 122–129, April 11–14, 1988.

[6] J.P. Lavieville, P. Carrere, T. Meynard, Electronic circuit for converting electrical energy and a power supply installation making use thereof. U.S. Patent 5 668 711 (1997).

Chapter 3

Advent of New Topologies

3.1 INTRODUCTION

A multilevel inverter (MLI) uses power semiconductor switches with several lower-voltage DC levels to perform DC-to-AC power conversion by synthesizing a staircase voltage waveform. Capacitors, batteries, and renewable energy voltage sources can be used as input DC levels. In general, the voltage stress on a power switch of an MLI is much lower than the operating voltage. Numerous advantages and some limitations of MLIs have already been discussed in earlier chapters.

Over the past few decades, MLIs have attracted wide interest both in the research community and in industry, as they are becoming a viable technology for many applications. In the mid-1970s, a patent describing the invention of a topology capable of producing multilevel voltage from separate DC sources was published by Baker and Bannister [1]. Since this topology comprised series-connected H-bridge inverters, it is known as the cascaded H-bridge (CHB) topology or series-connected H-bridge inverter. In another patent by Baker [2] in 1980, a modified multilevel topology was introduced. In contrast to the CHB inverters, this converter can synthesize multilevel waveforms from a single DC source, with additional diodes connected to the neutral point. This topology is now widely referred to as the neutral point-clamped (NPC) inverter and/or diode-clamped topology. In 1980, Nabae et al. [3] presented the implementation of an NPC inverter using a pulse width modulation scheme. In the 1980s much of the research was focused only on three-level inverters. The so-called flying-capacitor (FC) topology was introduced in the 1990s [4,5].

Many studies in the past few decades have shown an intense focus on the diode-clamped, FCs, and CHB topologies [6–36]. Thus, these three topologies are now widely referred to as "classical topologies." The recent research trend in MLIs in particular, and in power electronics in general, is the advent of new topologies. This trend pertains not only to MLIs but also to DC-to-DC, AC-to-DC, and AC-to-AC converters. This is mainly because of application requirements and advances in technologies related to semiconductor devices, sensors, modeling and simulation tools, and real-time processors. In this chapter, the advent of new topologies of MLIs is discussed. While specific examples of topologies are presented, you should be able to capture

Multilevel Inverters. DOI: https://doi.org/10.1016/B978-0-12-812448-2.00003-9

the systematic approach to analyze a given topology for its benefits and limitations.

3.2 ADVENT OF NEW TOPOLOGIES FOR MLIs

As discussed previously, with the emergence of MLIs, the so-called classical topologies attracted maximum attention both from academia and industry. Still, no specific topology can be said to be absolutely advantageous, because multilevel solutions are heavily dependent on application and cost considerations. A given topology can be very well adapted in some cases and totally unsuitable in others. Therefore, the optimal solution is often recommended on a case-by-case basis. Hence, along with the classical topologies, researchers continued (and still continue) to evolve newer topologies with an application-oriented approach. Some of these contributions are briefly summarized below. The aim is to enable the reader to appreciate that topological innovations are taking place in numerous ways and continue to be a hot trend.

Agelidis et al. [37] presented a single-phase voltage-source MLI topology for photovoltaic applications, which utilizes a combination of unidirectional and bidirectional switches of different ratings along with the use of a phase opposition carrier disposition multicarrier pulse-width modulation (PWM) switching technique.

A hybrid MLI based on CHB topology is presented for a 500-HP, 4.5-kV induction motor drive with investigations into design optimization, capacitor voltage balancing, and harmonic profile of the output waveform by Manjrekar and Lipo [38].

Xiaoming and Barbi [39] presented a modified diode-clamped topology to solve the problem of series-connected diodes in a classical diode-clamped inverter. In this structure, apart from the clamping of the main switches with clamping diodes, there is mutual clamping amongst the clamping diodes.

A generalized MLI was proposed by F.Z. Peng [40] with the primary objective of self-voltage-balancing, a capability which the classical inverters do not have. Although this topology needs many clamping switches, diodes, and capacitors, it presents a large number of redundant states and new topologies that can be further derived from it.

Cheng and Crow [41] have employed an additional circuitry integrated with the diode-clamped topology for implementing a static compensator with a battery energy storage system (STATCOM/BESS). This additional circuitry helps in effective balancing of the DC link capacitors.

A topology is proposed by Mariethoz and Rufer [42] by series combination of a three-phase six-switch voltage source inverter with single-phase H-bridges resulting in better waveform resolution and improved efficiency. This approach reduces the number of DC–DC converters supplying the cells of reversible multilevel converters.

A topology is proposed by Veenstra and Rufer [43] in which a three-phase three-level integrated gate-commutated thyristor inverter (as the main inverter) is connected in series with a two-level IGBT H-bridge (as the sub-inverter). Asymmetric source configuration is used to obtain a nine-level waveform for medium-voltage drive applications.

To obtain a multilevel voltage waveform in the event of partial failure(s) in the power circuit, a fault-tolerant topology was proposed by Chen et al. [44], which maintains the number of levels in the output voltage waveform with post-fault modification in the control signal, along with redundancy offered by multiswitching states.

Chen and He [45] have proposed a concept of using a mix of clamping diodes, power switches, and capacitors for a multilevel topology with self-voltage-balancing of the DC link capacitors for real and reactive power delivery.

To reduce the number of capacitors for a given number of levels, an approach using partial charging of capacitors was presented by Chan and Chau [46] by means of a new topology which also enables voltage boosting.

For the grid connection of two isolated photovoltaic generators with medium and high power ranges, a topology was proposed by Grandi et al. [47] using a dual two-level voltage source inverter so as to obtain a multilevel waveform. The approach reduces grid current harmonics.

Lezana et al. [48] have proposed an enhancement in the classical CHB topology with an active front end with the objective of a better regenerative mode of operation for loads which demand regenerative capability (such as laminators and downhill conveyors). This work demonstrates effective control of the input current and output voltage waveforms.

For standalone applications requiring a few kilowatts of power with a single battery storage, a topological solution has been presented by Daher et al. [49] with an MLI which uses a multiwinding transformer with appropriate turn ratios and an array of bidirectional power switches.

A diode-clamped based structure with a lower number of active devices was proposed by Gonzalez et al. [50] to eliminate the transformer from grid-connected photovoltaic power generation. Very interestingly, this topology simultaneously addresses consequent issues such as galvanic isolation safety requirements, leakage currents, and injection of DC into the grid.

For a transformerless multilevel structure, a topology was presented by Wang et al. [51], with floating capacitors forming series-connected independent cells with a large number of redundant states so as to achieve active voltage balancing without any auxiliary circuits.

Du et al. [52] have shown implementation of a cascaded multilevel boost inverter for electric vehicle and hybrid electric vehicle applications without the use of inductors.

Ewanchuk et al. [53] have proposed incorporation of coupled reactors in each leg of the NPC topology, for low-voltage high-speed motor applications, offering an increased number of output voltage levels, higher frequency waveforms, reduced dead-time effects, and a significant reduction in harmonics.

On the basis of the above glimpse into the evolution of topologies, it can be seen that, apart from exclusive and extensive studies into various aspects of the classical topologies, researchers continue to contribute towards new multilevel structures for different applications. Moreover, with the growing realization of the advantages and viability of MLIs, they are also being employed in low-power applications [54–57]. Hence, MLIs are receiving much more and wider attention, both in terms of topologies and control schemes. As discussed previously, multilevel topologies have an important limitation: they need a large number of power semiconductor switches, which increases the cost and control complexity and tends to reduce their overall reliability and efficiency. Although low-voltage-rated switches can be utilized in a multilevel converter, each switch requires a gate driver and protection circuit. This makes the overall system more expensive and complex. Consequently, for the last few years, efforts have been directed to reducing the device count in MLIs [58–76]. The focus of this chapter is on multilevel topologies with a reduction in the number of semiconductor devices. In the next section, a review of topologies is presented, introduced by their respective authors, with the sole objective of reducing the component count (primarily the number of power switches).

3.3 MLI TOPOLOGIES WITH REDUCED DEVICE COUNT

For the past few years, many researchers have been working on MLI topologies with a significant reduction in component count as compared to the classical topologies. Judging a topology without the consideration of the specific application is not only difficult, but also futile and unjustified. Often it can be seen that, while a topology is near-perfect for one application, it is useless for another. Still, in the context of this book, the general criteria for an overall assessment of a topology include:

1. the number of power switches used;
2. the total blocking voltage of the converter (which depends on the number of power switches and their respective voltage ratings);
3. the optimal controllability of the topology (in terms of possibility of charge-balance control and switching of the differently rated switches); and
4. possibility of employing asymmetric sources/capacitor voltage ratios.

While Parameters (1) and (2) directly influence the reliability of the inverter, efficiency is influenced by Parameters (1), (2), and (3) and application, performance and control complexity are governed by Parameter (3). The number of redundant states and consequently the programmability of fault-tolerant operation is directly influenced by (1) and (4). Based on these parameters, topologies with reduced component counts are discussed in this section.

3.3.1 Cascaded Half-Bridge-Based Multilevel DC Link Inverter

Gui-Jia [58,59] introduced a new class of MLIs based on a multilevel DC link (MLDCL) and a bridge inverter to reduce the device count. An MLDCL, as shown in Fig. 3.1, has cascaded half-bridge cells with each cell having its own DC source. A multilevel voltage-source inverter can be formed by connecting one of the MLDCLs with a single-phase bridge inverter. The MLDCL, which is comprised of sources $V_{DC,j}$ $\{j = 1-3\}$ and power switches S_j $\{j = 1-6\}$, generates a stepped waveform with one polarity, with or without PWM, to the bridge inverter, comprised of switches Q_j $\{j = 1-4\}$, which in turn alternates the polarity to produce an alternating voltage.

For an increased number of levels at the output, the MLDCL inverter can significantly reduce the switch count as well as the number of gate drivers as compared to the CHB topology [58]. With reference to Fig. 3.1, various combinations that can be obtained for the MLDCL v_{bus} are summarized in Table 3.1.

It can be observed that, to obtain a given level, three switches conduct simultaneously in the MLDCL part and two switches conduct in the H-bridge part (Switches Q_1 and Q_4 for the positive half cycle, Q_2 and Q_3 for the negative half cycle, and Q_1 and Q_3 or Q_2 and Q_4 for zero level). It can

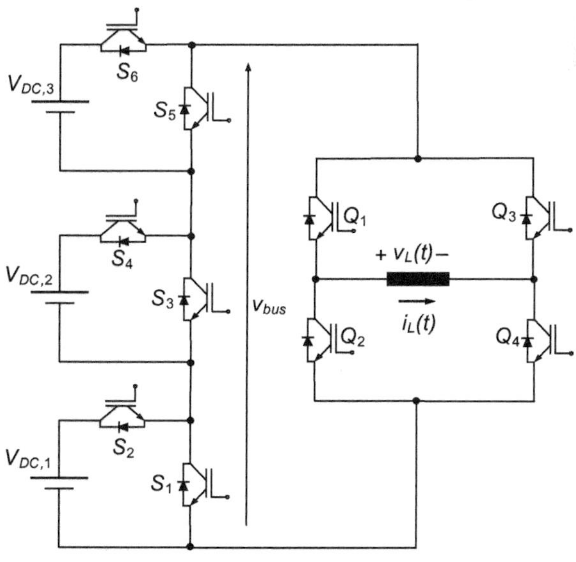

FIGURE 3.1 MLDCL inverter.

TABLE 3.1 Switching States for the Topology Shown in Fig. 3.1

State	Multilevel DC Link Bus Voltage v_{bus}	Switches in ON State
1	$V_{DC,1}$	S_2, S_3, S_5
2	$V_{DC,2}$	S_1, S_4, S_5
3	$V_{DC,3}$	S_1, S_3, S_6
4	$V_{DC,1} + V_{DC,2}$	S_2, S_4, S_5
5	$V_{DC,2} + V_{DC,3}$	S_1, S_4, S_6
6	$V_{DC,1} + V_{DC,3}$	S_2, S_3, S_6
7	$V_{DC,1} + V_{DC,2} + V_{DC,3}$	S_2, S_4, S_6
8	0	S_1, S_3, S_5

be observed from the topology that each power switch of the H-bridge part must have a minimum voltage-blocking capability equal to the sum of the input voltage values. Thus, these switches are rated higher compared to the switches in the MLDCL part. However, since the zero level can be synthesized using switches of the MLDCL part, the higher-rated switches Q_j $\{j = 1-4\}$ can be operated at fundamental switching frequency.

For a symmetric source configuration with $V_{DC,1} = V_{DC,2} = V_{DC,3} = V_{DC}$, it can be seen that the switches S_j $\{j = 1-6\}$ need to block a voltage of V_{DC} and need to conduct a current equal to the load current. Switches Q_j $\{j = 1-4\}$ need to block a voltage equal to $3V_{DC}$ and conduct a current equal to the load current. It can also be observed from Table 3.1 that since voltage levels V_{DC}, $2V_{DC}$, and $3V_{DC}$ can be synthesized combining all the input sources in groups of one, two, and three, respectively, equal load sharing amongst them is possible. These redundancies also provide flexibility in voltage balancing when capacitors are used.

Regarding asymmetric source configurations in MLDCL topology, no analysis is offered in References [58,59]. Since subtractive combinations of the input DC levels cannot be synthesized, the trinary source configuration (i.e., $V_{DC,j} = 3^{(j-1)} V_{DC}$) cannot be employed for this topology. As can be seen from Table 3.1, a binary combination with $V_{DC,1} = V_{DC}$, $V_{DC,2} = 2V_{DC}$ and $V_{DC,3} = 4V_{DC}$ is possible since the voltage levels V_{DC}, $2V_{DC}$, $3V_{DC}$, $4V_{DC}$, $5V_{DC}$, $6V_{DC}$, and $7V_{DC}$ can be synthesized by using States 1, 2, 4, 3, 6, 5, and 7, respectively.

3.3.2 T-Type Inverter

Gerardo Ceglia et al. [60–62] have presented a new MLI topology, named the T-type inverter, comprising an H-bridge output stage and bidirectional auxiliary switches. It offers a significant reduction in the number of power devices as compared to conventional topologies. A single-phase structure with four input voltage sources, namely, $V_{DC,1}$, $V_{DC,2}$, $V_{DC,3}$, and $V_{DC,4}$, is shown in Fig. 3.2. It comprises three switches S_j $\{j = 1-3\}$ which are bidirectional-blocking-bidirectional-conducting, while four switches Q_j $\{j = 1-4\}$ are unidirectional-blocking-bidirectional-conducting.

Thus, this topology inadvertently requires a mix of unidirectional and bidirectional power switches. The valid switching states are summarized in Table 3.2 and it can be seen that the input DC values are required to be symmetric, i.e., $V_{DC,1} = V_{DC,2} = V_{DC,3} = V_{DC,4} = V_{DC}$. This is because not all combinations of input voltage levels can be synthesized at the load terminals. In most cases, either a positive or negative combination can be synthesized, but not both.

For example, while a voltage level $-V_{DC,4}$ can be synthesized at the load terminals, the level $+V_{DC,4}$ cannot be synthesized. Thus, it is essential that the input sources are symmetric. Also, lack of sufficient redundancies goes against effective voltage balancing. It can be also observed from Table 3.2 that equal load sharing amongst the input voltage sources is not possible as the number of valid states is very limited. For a given state, only two switches conduct simultaneously. The bidirectional switches are voltage-rated at different values. While S_3 should be minimally rated at $3V_{DC}$, S_2 and S_1 should be rated at $2V_{DC}$ each. The H-bridge switches must each have minimum blocking capability of $4V_{DC}$. These higher voltage switches, however, can be operated with fundamental switching frequency.

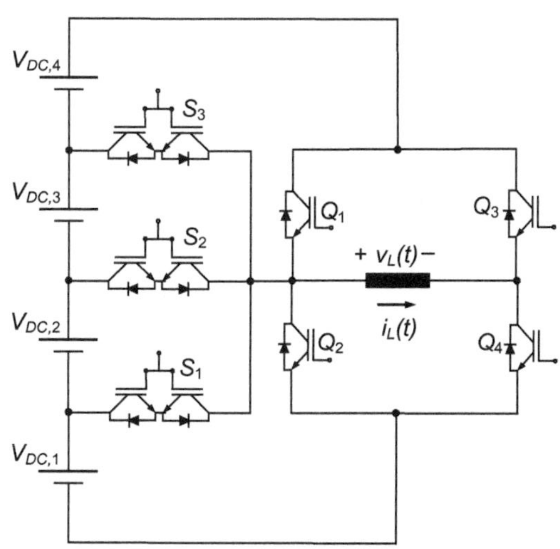

FIGURE 3.2 A T-type inverter with four input sources.

TABLE 3.2 Possibilities of Switching Combinations for the Topology Shown in Fig. 3.2

State	Output Voltage $v_L(t)$	Switches in ON State (Other Switches Remain OFF)
1	$+V_{DC,1}$	S_1, Q_4
2	$-V_{DC,1}$	Not possible
3	$+V_{DC,2}$	Not possible
4	$-V_{DC,2}$	Not possible
5	$+V_{DC,3}$	Not possible
6	$-V_{DC,3}$	Not possible
7	$+V_{DC,4}$	Not possible
8	$-V_{DC,4}$	S_3, Q_3
9	$V_{DC,1}+V_{DC,2}$	S_2, Q_4
10	$-(V_{DC,1}+V_{DC,2})$	Not possible
11	$V_{DC,2}+V_{DC,3}$	Not possible
12	$-(V_{DC,2}+V_{DC,3})$	Not possible
13	$V_{DC,3}+V_{DC,4}$	Not possible
14	$-(V_{DC,3}+V_{DC,4})$	S_2, Q_3
15	$V_{DC,1}+V_{DC,3}$	Not possible
16	$-(V_{DC,1}+V_{DC,3})$	Not possible
17	$V_{DC,1}+V_{DC,4}$	Not possible
18	$-(V_{DC,1}+V_{DC,4})$	Not possible
19	$V_{DC,2}+V_{DC,4}$	Not possible
20	$-(V_{DC,2}+V_{DC,4})$	Not possible
21	$V_{DC,1}+V_{DC,2}+V_{DC,3}$	S_3, Q_4
22	$-(V_{DC,1}+V_{DC,2}+V_{DC,3})$	Not possible
23	$V_{DC,2}+V_{DC,3}+V_{DC,4}$	Not possible
24	$-(V_{DC,2}+V_{DC,3}+V_{DC,4})$	S_1, Q_3
25	$V_{DC,1}+V_{DC,2}+V_{DC,4}$	Not possible
26	$-(V_{DC,1}+V_{DC,2}+V_{DC,4})$	Not possible
27	$V_{DC,1}+V_{DC,3}+V_{DC,4}$	Not possible
28	$-(V_{DC,1}+V_{DC,3}+V_{DC,4})$	Not possible

(Continued)

TABLE 3.2 (Continued)

State	Output Voltage $v_L(t)$	Switches in ON State (Other Switches Remain OFF)
29	$V_{DC,1} + V_{DC,2} + V_{DC,3} + V_{DC,4}$	Q_1, Q_4
30	$-(V_{DC,1} + V_{DC,2} + V_{DC,3} + V_{DC,4})$	Q_2, Q_3
31	0	Q_1, Q_3
32	0	Q_2, Q_4

3.3.3 Switched Series/Parallel Sources–Based MLI

Hinago and Koizumi [63,64] have proposed an MLI topology consisting of an H-bridge and DC sources which can be switched in series and in parallel. The topology reduces the number of gate driver circuits, thereby reducing the size and power consumption. The topology synthesizes the same number of output levels with a lower number of power switches as compared to a CHB topology. An important suggested application is for electric vehicular applications where a single battery composed of a number of series-connected battery cells is available, which can be rearranged using the switched sources topology, hence reducing the requirement for switching devices. More importantly, the possibility of combining two or more sources in series and parallel gives enough flexibility for meeting voltage/power requirements in the vehicle drive system.

An Switched Series/Parallel Sources–MLI with three input DC sources is shown in Fig. 3.3. It consists of two parts: the switched sources part which synthesizes a bus voltage v_{bus} *(t)* and the H-bridge part which synthesizes positive and negative cycles of voltage v_{bus} *(t)* to feed an AC load. Three sources, namely, $V_{DC,1}$, $V_{DC,2}$, and $V_{DC,3}$ and power switches S_j $\{j = 1-6\}$ constitute the switched sources part. Power switches Q_j $\{j = 1-4\}$ constitute the H-bridge part. Levels that can be synthesized by the switched sources part are summarized in Table 3.3.

For a symmetric source configuration, i.e., $V_{DC,1} = V_{DC,2} = V_{DC,3} = V_{DC}$, it can be seen from Table 3.3 that voltage levels V_{DC} and $2V_{DC}$ can be synthesized with three states each, while one state is available for voltage level $3V_{DC}$. Moreover, the voltage stress experienced by the switches S_j $\{j = 1-6\}$ would each be equal to V_{DC}. An important limitation of this topology is that the switches Q_j $\{j = 1-4\}$ need to have a minimum blocking capability of summation of voltages of all voltage sources. Thus, for the symmetric source configuration with three sources, the H-bridge switches should possess a voltage-blocking capability of $3V_{DC}$. Another important limitation is that these switches with higher blocking capability cannot be operated at

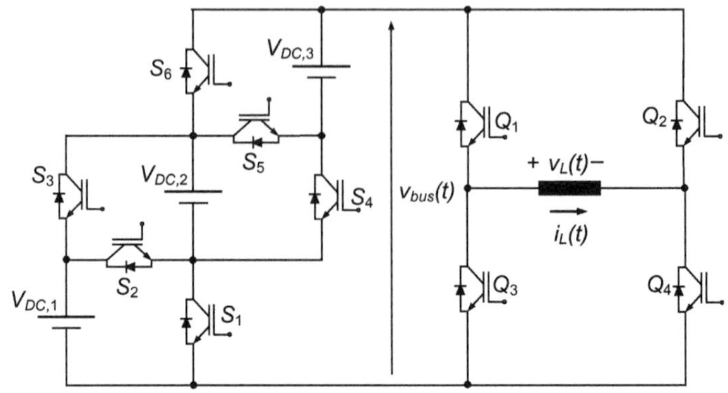

FIGURE 3.3 MLI topology as proposed in References [63,64].

TABLE 3.3 Switching States for Topology Shown in Fig. 3.3

State	Bus Voltage $v_{bus}(t)$	Switches in ON State
1	$V_{DC,1}$	S_3, S_6
2	$V_{DC,2}$	S_1, S_6
3	$V_{DC,3}$	S_1, S_4
4	$V_{DC,1} + V_{DC,2}$	S_2, S_6
5	$V_{DC,2} + V_{DC,3}$	S_1, S_5
6	$V_{DC,1} + V_{DC,3}$	S_3, S_5
7	$V_{DC,1} + V_{DC,2} + V_{DC,3}$	S_2, S_5

fundamental switching frequencies because the zero voltage level is not synthesized by the switched sources part, as can be observed from Table 3.3. It can also be inferred from the table that, with input sources of equal voltages, equal load sharing amongst them is possible as the sources can be combined in all additive configurations. Various combinations can be utilized in different cycles so as to equate the average current from each source, thereby equalizing the average power amongst the sources.

Although the topology enables the synthesis of all additive combinations of the input sources, subtractive combinations are not possible. Hence, trinary source configuration is not possible for this topology. Binary source configuration is, however, possible so as to maximize the number of levels in the output waveform. For example, in Fig. 3.3, for $V_{DC,1} = V_{DC}$, $V_{DC,2} = 2V_{DC}$, and $V_{DC,3} = 4V_{DC}$, all possible combinations, namely, V_{DC}, $2V_{DC}$, $3V_{DC}$, $4V_{DC}$, $5V_{DC}$, $6V_{DC}$, and $7V_{DC}$ are obtained as

$v_{bus}(t)$ by using States 1, 2, 4, 3, 6, 5, and 7, so that the load voltage waveform has 15 levels in equal steps of V_{DC}. Thus, binary, but not trinary, source configuration can be implemented using this topology.

3.3.4 Series-Connected Switched Sources–Based MLI

A modified form of CHB topology is presented in References [65,66], with the objective of reducing the number of switches as compared to the classical CHB topology. The topology with four input DC sources $V_{DC,j}$ $\{j = 1-4\}$ is shown in Fig. 3.4. The low potential terminals of the sources are all connected through power switches. These terminals are also connected to the higher potential terminal of the preceding source through power switches, as illustrated in Fig. 3.4 with S_j $\{j = 1-8\}$. This interconnection is capable of synthesizing a multilevel rectified waveform v_{bus} which is imparted positive and negative polarities using the H-bridge comprising switches Q_j $\{j = 1-4\}$.

The possibilities of synthesizing various combinations of input DC levels are summarized in Table 3.4. It can be seen that the structure, though simple, allows a much lower number of valid states. In fact, not even the input DC levels offered by the sources can all be obtained as v_{bus}, except that of $V_{DC,1}$.

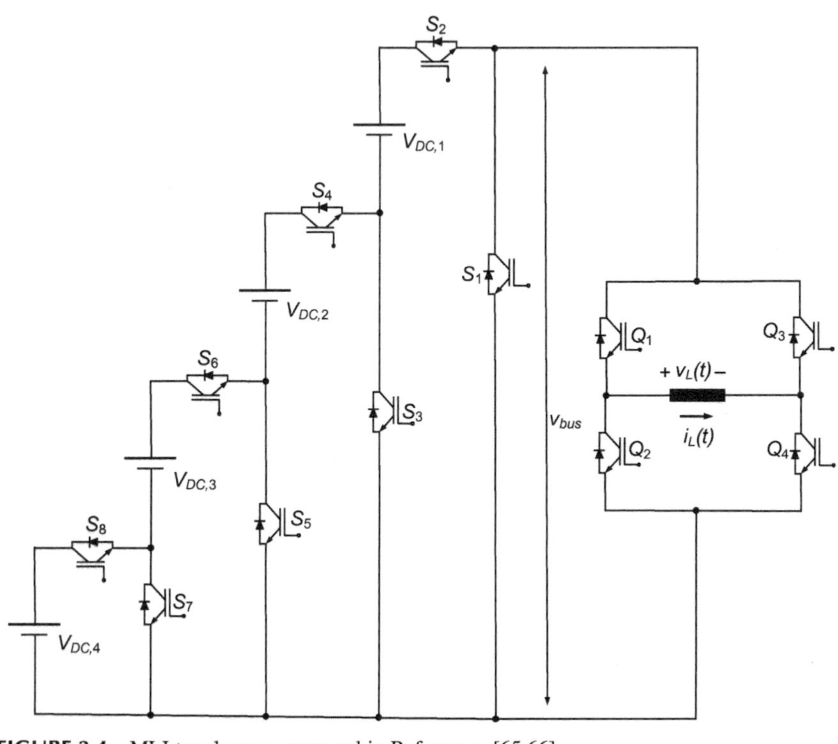

FIGURE 3.4 MLI topology as proposed in References [65,66].

TABLE 3.4 Possibilities of Switching Combinations for the Topology Shown in Fig. 3.4

State	Voltage Level v_{bus}	Switches in ON State
1	$V_{DC,1}$	S_1
2	$V_{DC,2}$	Not possible
3	$V_{DC,3}$	Not possible
4	$V_{DC,4}$	Not possible
5	$V_{DC,1} + V_{DC,2}$	S_2, S_3
6	$V_{DC,2} + V_{DC,3}$	Not possible
7	$V_{DC,3} + V_{DC,4}$	Not possible
8	$V_{DC,1} + V_{DC,4}$	Not possible
9	$V_{DC,2} + V_{DC,4}$	Not possible
10	$V_{DC,1} + V_{DC,3}$	Not possible
11	$V_{DC,1} + V_{DC,2} + V_{DC,3}$	S_2, S_4, S_5
12	$V_{DC,2} + V_{DC,3} + V_{DC,4}$	Not possible
13	$V_{DC,1} + V_{DC,3} + V_{DC,4}$	Not possible
14	$V_{DC,1} + V_{DC,2} + V_{DC,4}$	Not possible
15	$V_{DC,1} + V_{DC,2} + V_{DC,3} + V_{DC,4}$	S_2, S_4, S_6
16	0	S_1

Thus, this topology does not offer any possibility of employing asymmetric source configurations. The source configuration mandatorily needs to be symmetric, i.e., $V_{DC,1} = V_{DC,2} = V_{DC,3} = V_{DC,4} = V_{DC}$. With such a configuration, various switches would be differently voltage-rated. Switches Q_j $\{j = 1-4\}$ should be minimally rated at $4V_{DC}$, S_1 should be rated minimally at $3V_{DC}$, S_3 should be rated at $2V_{DC}$, and S_2, S_4, S_5, and S_6 should be minimally rated at V_{DC} each. Moreover, as can be seen from Table 3.4, for symmetric input sources, equal load sharing is not possible as there are many combinations of input DC levels which are not feasible. Also, since the zero level can be obtained as v_{bus}, the higher-rated switches Q_j $\{j = 1-4\}$ can be operated at fundamental switching frequency.

3.3.5 Cascaded "Bipolar Switches Cells"–Based MLI

Babaei et al. [67] have presented a new class of multilevel converters to reduce the number of power electronics switches and DC voltage sources.

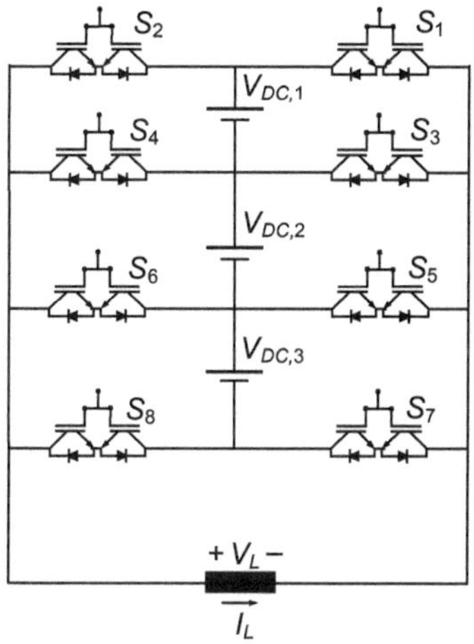

FIGURE 3.5 MLI topology as proposed in Reference [67].

Fig. 3.5 shows the single-phase structure of a seven-level converter with three input voltage sources with equal voltages of V_{DC} (symmetrical source configuration). All the switches are bidirectional-blocking-bidirectional-conducting. Although each bidirectional switch requires two IGBTs, the total number of gate drive circuits is equal to the number of bidirectional switches. This results in reducing the cost and overall complexity of the converter.

Switching states for all possible combinations of input voltage sources are shown in Table 3.5. It can be observed that while synthesizing $2V_{DC}$ and $-2V_{DC}$, not all possible combinations of input voltage sources are utilized. As a result, equal utilization of the input voltage sources is not possible in this topology. Moreover, the outermost bidirectional switches S_1, S_2, S_7, and S_8 need to have minimum voltage blocking capability of $3V_{DC}$ each. On the other hand, the inner switches S_3, S_4, S_5, and S_6 need to have minimum voltage blocking capability of $2V_{DC}$. It can also be observed that to synthesize each voltage level, only two switches need to conduct simultaneously. It should also be noted that the topology can only work with symmetrical sources configuration.

3.3.6 Packed U-Cell Topology

Youssef Ounejjar et al. [68–72] have proposed a new topology which is very competitive compared to the classical topologies. It consists of the so-called "packed U cells." Each U cell consists of an arrangement of two power switches and one DC input level (obtained with a voltage source or a floating capacitor). A single-phase structure of the packed U-cell (PUC) topology with two input DC levels, namely, $V_{DC,1}$ and $V_{DC,2}$, and six switches S_j $\{j = 1-6\}$, is shown in Fig. 3.6.

TABLE 3.5 Possibilities of Switching Combinations for the Topology Shown in Fig. 3.5

State	Output Voltage $V_L(t)$	Switches in ON State (Other Switches Remain OFF)	Remarks
1	$V_{DC,1}$	S_2, S_3	
2	$V_{DC,2}$	S_4, S_5	
3	$V_{DC,3}$	S_6, S_7	
4	$V_{DC,1} + V_{DC,2}$	S_2, S_5	
5	$V_{DC,2} + V_{DC,3}$	S_4, S_7	
6	$V_{DC,3} + V_{DC,1}$	S_2, S_3, S_5, S_7	Not possible as $V_{DC,2}$ gets short-circuited
7	$V_{DC,1} + V_{DC,2} + V_{DC,3}$	S_2, S_7	
8	$-V_{DC,1}$	S_1, S_4	
9	$-V_{DC,2}$	S_3, S_6	
10	$-V_{DC,3}$	S_5, S_8	
11	$-(V_{DC,1} + V_{DC,2})$	S_1, S_6	
12	$-(V_{DC,2} + V_{DC,3})$	S_3, S_8	
13	$-(V_{DC,3} + V_{DC,1})$	S_1, S_4, S_6, S_8	Not possible as $V_{DC,2}$ gets short-circuited
14	$-(V_{DC,1} + V_{DC,2} + V_{DC,3})$	S_1, S_8	
15	0	S_7, S_8	

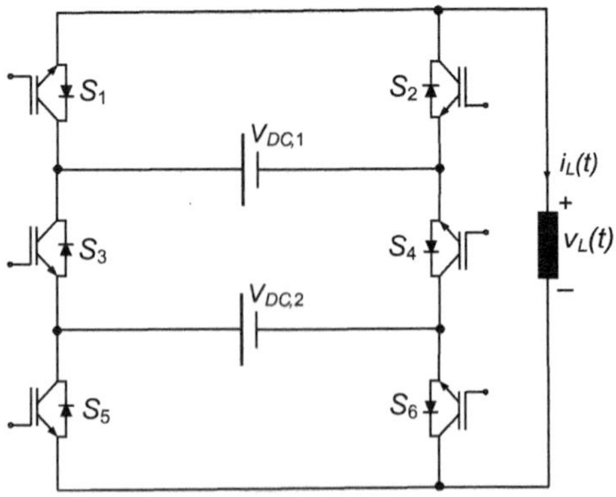

FIGURE 3.6 PUC MLI topology as proposed in References [68–72].

TABLE 3.6 Switching States for Topology Shown in Fig. 3.6

State	Output Voltage $v_L(t)$	Switches in ON State (Other Switches Remain OFF)
1	0	S_1, S_3, S_5
2	0	S_2, S_4, S_6
3	$V_{DC,1}$	S_1, S_4, S_6
4	$V_{DC,2}$	S_1, S_3, S_6
5	$V_{DC,1} - V_{DC,2}$	S_1, S_4, S_5
6	$-V_{DC,1}$	S_2, S_3, S_5
7	$-V_{DC,2}$	S_2, S_4, S_5
8	$-V_{DC,1} + V_{DC,2}$	S_2, S_3, S_6

The topology is very simple in terms of the interconnection of components. The minimum voltage-blocking capability required for switches (S_1, S_2) is $V_{DC,1}$, while that for (S_5, S_6) is $V_{DC,2}$. The voltage-blocking capability for switches (S_3, S_4) is ($V_{DC,1}-V_{DC,2}$). Various states for the structure are shown in Table 3.6. Thus, to obtain a desired voltage level, only three switches conduct simultaneously. It is important to observe from Table 3.6 that to derive the desired benefit from the topology, symmetric source configuration cannot be used as the output will be three-level with many redundant states.

An elaborate methodology is presented in Reference [72] to calculate the asymmetric voltage levels. For two input levels, the number of output levels can be maximized with $V_{DC,1} = 3V_{DC}$ and $V_{DC,2} = V_{DC}$. This trinary configuration synthesizes seven levels, not nine, because of the absence of additive combination. For a structure with two input sources, switching of the middle two switches, namely (S_3, S_4), can be performed at a fundamental frequency as demonstrated in Reference [72]. This feature, however, is not feasible for PUC-based structures with more than two input DC levels.

3.3.7 Multilevel Module–Based MLI

The topology presented in Reference [73] also aims to reduce the number of switches. This topology is a hybrid multilevel topology comprising of two parts. The "level generation part" consists of input DC sources and bidirectional-blocking-bidirectional-conducting switches and it synthesizes a multilevel half-wave voltage waveform. The voltage stress on these switches is not distributed uniformly. The "polarity generation part" is a simple H-bridge connected at the output of the level generation part which generates

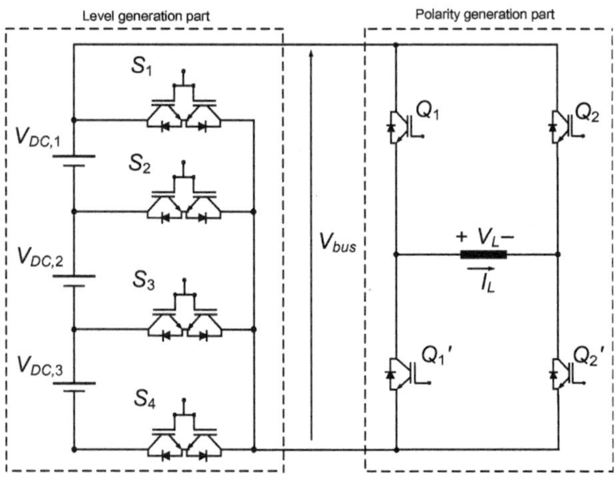

FIGURE 3.7 MLI topology as proposed in Reference [73].

negative and positive polarities for the output voltage. The switches in this part are unidirectional-blocking-bidirectional-conducting and have to withstand the maximum voltage generated by the level generation part. However, these switches can be operated at line frequency. Thus these switches are high-voltage low-frequency switches.

A single-phase topology for a seven-level converter is shown in Fig. 3.7, where $V_{DC,1} = V_{DC,2} = V_{DC,3} = V_{DC}$. The operating states are listed in Table 3.7. It is evident that all possible combinations of the input voltage levels are not utilized. Thus, in this topology, equal load sharing among the input DC sources is not possible.

The switches in the H-bridge, Q_1, Q_2, Q_1', and Q_2', are each subjected to the voltage stress of $3V_{DC}$. Switches S_1 and S_4 need to have a minimum voltage-blocking capability of $3V_{DC}$, whereas switches S_2 and S_3 should be selected to bear the voltage stress of $2V_{DC}$. However, only one switch in the level generation part and two switches in the polarity generation part need to conduct simultaneously. The topology does not allow asymmetrical source configuration (binary or trinary), because it is not possible to synthesize all subtractive and additive combinations.

3.3.8 Reversing Voltage Topology

The reversing voltage topology proposed in References [74,75] is illustrated in Fig. 3.8. It is a hybrid topology which separates the output voltage into two parts. One part synthesizes a multilevel stepped voltage half-wave. This part is named the "level generation part," comprising input DC sources $V_{DC,\,J}$ $\{j = 1{-}4\}$ and switches S_j $\{j = 1{-}8\}$. A full-bridge, comprising switches Q_j $\{j = 1{-}8\}$, is connected to the level generation part so as to obtain both

TABLE 3.7 Switching Combinations for the Topology Shown in Fig. 3.7

State	Output Voltage at Level Generation Part $V_{bus}(t)$	Switches in ON State (Other Switches Remain OFF)	Remarks
1	$V_{DC,1}$	S_2	
2	$V_{DC,2}$	S_1, S_2, S_3	Not possible as both $V_{DC,1}$ and $V_{DC,2}$ get short-circuited
3	$V_{DC,3}$	S_1, S_3, S_4	Not possible as $V_{DC,3}$ gets short-circuited
4	$V_{DC,1} + V_{DC,2}$	S_3	
5	$V_{DC,2} + V_{DC,3}$	S_1, S_2, S_4	Not possible as $V_{DC,1}$ gets short-circuited
6	$V_{DC,3} + V_{DC,1}$	S_2, S_3, S_4	Not possible as both $V_{DC,2}$ and $V_{DC,3}$ get short-circuited
7	$V_{DC,1} + V_{DC,2} + V_{DC,3}$	S_4	
8	0	S_1	

positive and negative polarities for the output voltage. This part is named the "polarity generation part." In this way, the components are utilized effectively. The switches in the polarity generation part need to withstand the total additive voltage of the level generation part. This topology exhibits modularity for the level generation part.

Various possible states to obtain different levels at the level generation part, $v_{bus}(t)$, are summarized in Table 3.8. It can be noted that the switches with high blocking voltages, Q_j $\{j = 1-4\}$, can be operated at fundamental switching frequency. If symmetric sources are used such that $V_{DC,1} = V_{DC,2} = V_{DC,3} = V_{DC,4} = V_{DC}$, then all switches of the level generation part experience a voltage stress of V_{DC}, while the four switches of the polarity generation part are required to have a minimum voltage-blocking capability of $4V_{DC}$ each. It can also be inferred from Table 3.3 that for symmetric source configuration, equal load sharing amongst them is not possible. For a DC link created with connected capacitors, this limitation will affect voltage balancing in the capacitors. It can also be observed from Table 3.8 that the number of switches conducting simultaneously to synthesize a required voltage level is not the same, and thus conduction losses and switching losses for the switches may not be the same. Moreover, since the topology

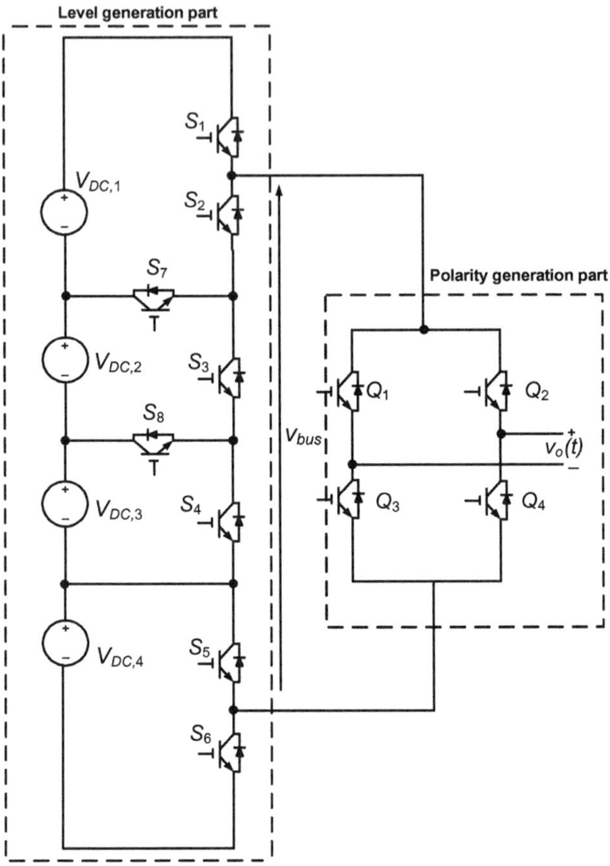

FIGURE 3.8 MLI topology as proposed in References [74,75].

does not facilitate the synthesis of all additive and subtractive combinations of input voltage sources, trinary source combinations cannot be used.

Employing other asymmetric combinations to maximize the number of output levels is seriously hampered by the absence of some states with a single voltage source (e.g., as seen from Table 3.8, using State 2 for engaging the source $V_{DC,2}$ would lead to short-circuiting of source $V_{DC,3}$ and hence this state cannot be used). However, one important advantage that the topology offers is that it uses a single DC link for three-phase implementation.

3.3.9 Two-Switch Enabled Level Generation—Based MLI

The topology presented by Babaei [76] requires a mix of unidirectional and bidirectional switches, as shown in Fig. 3.9. It also consists of two parts, wherein one part helps in the generation of voltage levels and the other takes

TABLE 3.8 Possible Switching Combinations for the Topology Shown in Fig. 3.8

State	Output Voltage at Level Generation Part $v_{bus}(t)$	Switches in ON State (Other Switches Remain OFF)	Remarks
1	$V_{DC,1}$	S_1, S_3, S_4, S_5, S_7	–
2	$V_{DC,2}$	S_2, S_4, S_5, S_7, S_8	Not possible as $V_{DC,3}$ gets short-circuited
3	$V_{DC,3}$	S_2, S_3, S_5, S_8	–
4	$V_{DC,4}$	S_2, S_3, S_4, S_6	–
5	$V_{DC,1}$ + $V_{DC,2}$	S_1, S_4, S_5, S_8	–
6	$V_{DC,2}$ + $V_{DC,3}$	S_2, S_5, S_7	–
7	$V_{DC,3}$ + $V_{DC,4}$	S_2, S_3, S_6, S_8	–
8	$V_{DC,1}$ + $V_{DC,4}$	S_1, S_3, S_4, S_6, S_7	–
9	$V_{DC,2}$ + $V_{DC,4}$	S_2, S_4, S_6, S_7, S_8	Not possible as $V_{DC,3}$ gets short-circuited
10	$V_{DC,1}$ + $V_{DC,3}$	S_1, S_3, S_5, S_7, S_8	Not possible as $V_{DC,2}$ gets short-circuited
11	$V_{DC,1}$ + $V_{DC,2}$ + $V_{DC,3}$	S_1, S_5	–
12	$V_{DC,2}$ + $V_{DC,3}$ + $V_{DC,4}$	S_2, S_6, S_7	–
13	$V_{DC,1}$ + $V_{DC,3}$ + $V_{DC,4}$	S_1, S_3, S_6, S_7, S_8	Not possible as $V_{DC,2}$ gets short-circuited
14	$V_{DC,1}$ + $V_{DC,2}$ + $V_{DC,4}$	S_1, S_4, S_6, S_8	Not possible as $V_{DC,3}$ gets short-circuited
15	$V_{DC,1}$ + $V_{DC,2}$ + $V_{DC,3}$ + $V_{DC,4}$	S_1, S_6	–
16	0	S_2, S_3, S_4, S_5	–

part in the generation of polarities. The level generation part requires both unidirectional and bidirectional switches. However, this part is unable to synthesize the zero level. The polarity generation part is an H-bridge with unidirectional switches. These switches operate under the voltage stress

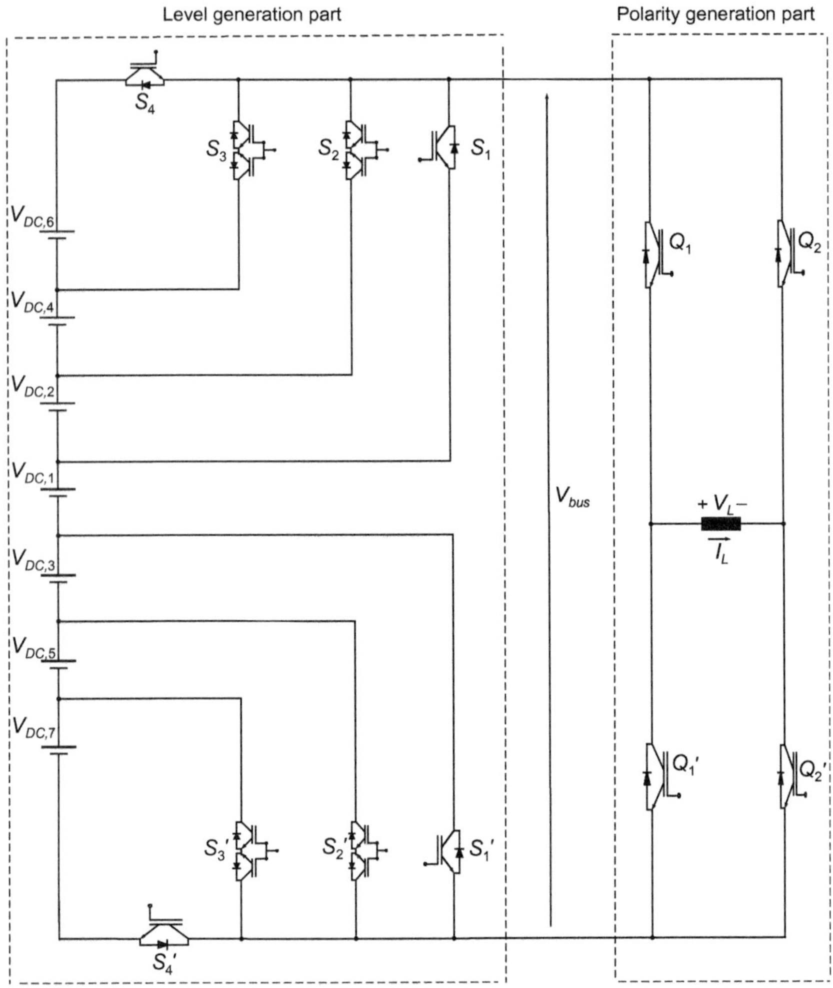

FIGURE 3.9 MLI topology as proposed in Reference [76].

equal to the maximum voltage generated at $V_{bus}(t)$. Since the level generation part is not able to synthesize the zero level, the switches in the H-bridge cannot be operated at line frequency.

A single-phase structure with seven input DC voltage sources is shown in Fig. 3.9. For a symmetrical source configuration, $V_{DC,1} = V_{DC,2} = V_{DC,3} = V_{DC,4} = V_{DC,5} = V_{DC,6} = V_{DC,7} = V_{DC}$. Table 3.9 lists the various operating states for the given structure. It can be observed that only two switches are conducting per state in the level generation part, and hence it is called "two-switch enabled level generation-based MLI." The zero level is synthesized using the H-bridge. It is not possible to apply the concept of charge balancing in this topology as all the sources do not contribute equally for each level

TABLE 3.9 Switching States for Topology Shown in Fig. 3.9

State	Output Voltage at Level Generation Part $V_{bus}(t)$	Switches in ON State (Other Switches Remain OFF)
1	$V_{DC,1}$	S_1, $S_1^{'}$
2	$V_{DC,1} + V_{DC,2}$	S_2, $S_1^{'}$
3	$V_{DC,1} + V_{DC,2} + V_{DC,3}$	S_2, $S_2^{'}$
4	$V_{DC,1} + V_{DC,2} + V_{DC,3} + V_{DC,4}$	S_3, $S_2^{'}$
5	$V_{DC,1} + V_{DC,2} + V_{DC,3} + V_{DC,4} + V_{DC,5}$	S_3, $S_3^{'}$
6	$V_{DC,1} + V_{DC,2} + V_{DC,3} + V_{DC,4} + V_{DC,5} + V_{DC,6}$	S_4, $S_3^{'}$
7	$V_{DC,1} + V_{DC,2} + V_{DC,3} + V_{DC,4} + V_{DC,5} + V_{DC,6} + V_{DC,7}$	S_4, $S_4^{'}$

in the $V_{bus}(t)$. The switches in the H-bridge, Q_1, $Q_2^{'}$, $Q_1^{'}$, and Q_2, need to have a minimum voltage-blocking capability of $7V_{DC}$. Switches S_1, S_4, $S_1^{'}$, and $S_4^{'}$, of the level generation part need to have a minimum voltage-blocking capability of $3V_{DC}$. The remainder of the switches need to have a minimum voltage-blocking capability of $2V_{DC}$ .It is also observed that this topology does not support asymmetrical source configuration (binary or trinary) as it is not possible to synthesize all subtractive and additive combinations of the input voltage levels. However, one advantage that this topology offers is that a total of four power electronic switches need to be conducting in any state, resulting in lower conduction losses.

Thus, the description of topologies proposed for reducing the component count indicates that a given structure has its pros and cons.

3.4 SUMMARY

As MLIs continue to gain increasing importance for both high-power and low-power applications, many researchers have proposed specific topological solutions for intended applications. Also, new multilevel topologies have been proposed offering high output resolution with a reduced number of devices. Based on the discussion on new topologies, it can be concluded that in the process of reducing the device count, various compromises are involved which affect the topology characteristics, such as:

1. the possibility of charge balance control (equal load sharing amongst the input sources);
2. optimal switching of the differently rated power switches; and
3. the possibility of asymmetric source configuration.

In this book, topological solutions will be further explored which can preserve the desired characteristics, while reducing the device count for voltage output with a large number of levels. In addition, a comparison of these topologies will be carried out with the classical topologies in terms of the number of power switches, isolated sources/capacitors, clamping components, voltage stresses, etc.

REFERENCES

[1] R.H. Baker, L.H. Bannister, Electric power converter. U.S. Patent 3 867 643, 1975.

[2] R.H. Baker Switching circuit. U.S.Patent 4210826, 1980.

[3] A. Nabae, I. Takahashi, H. Akagi, A neutral-point-clamped PWM inverter, in: Conference Record IEEE IAS Annual Meeting, Cincinnati, OH, vol. 3, pp. 761–766, September 28–October 3, 1980.

[4] M. Marchesoni, M. Mazzucchelli, S. Tenconi, A non conventional power converter for plasma stabilization, Proceedings of the IEEE 19th Power Electronics Specialist Conference (1988) 122–129. April 11–14.

[5] J.P. Lavieville, P. Carrere, T. Meynard, Electronic circuit for converting electrical energy and a power supply installation making use thereof, U.S. Patent 5 668 711, 1997.

[6] J.A. Barrena, L. Marroyo, M.A.R. Vidal, J.R.T Apraiz, Individual voltage balancing strategy for PWM cascaded H-bridge converter-based STATCOM, IEEE Trans. Industrial Electron. 55 (1) (2008) 21–29.

[7] D. Zhong, L.M. Tolbert, B. Ozpineci, J.N. Chiasson, Fundamental frequency switching strategies of a seven-level hybrid cascaded H-bridge multilevel inverter, IEEE Trans. Pow. Electron. 24 (1) (2009) 25–33.

[8] R. Teodorescu, F. Blaabjerg, J.K. Pedersen, E. Cengelci, P.N. Enjeti, Multilevel inverter by cascading industrial VSI, IEEE Trans. Industrial Electron. 49 (4) (2002) 832–838.

[9] M.A. Perez, P. Cortes, J. Rodriguez, Predictive control algorithm technique for multilevel asymmetric cascaded H-bridge inverters, IEEE Trans. Industrial Electron. 55 (12) (2008) 4354–4361.

[10] A.K. Gupta, A.M. Khambadkone, A space vector modulation scheme to reduce common mode voltage for cascaded multilevel inverters, IEEE Trans. Pow. Electron. 22 (5) (2007) 1672–1681.

[11] S. Vazquez, J.I. Leon, J.M. Carrasco, L.G. Franquelo, E. Galvan, M. Reyes, J.A. Sanchez, E. Dominguez, Analysis of the power balance in the cells of a multilevel cascaded H-bridge converter, IEEE Trans. Industrial Electron. 57 (7) (2010) 2287–2296.

[12] P. Barriuso, J. Dixon, P. Flores, L. Moran, Fault-tolerant reconfiguration system for asymmetric multilevel converters using bidirectional power switches, IEEE Trans. Industrial Electron. 56 (4) (2009) 1300–1306.

[13] Y. Li, B. Wu, A novel DC voltage detection technique in the CHB inverter-based STATCOM, IEEE Trans. Pow. Deliv. 23 (3) (2008) 1613–1619.

[14] P. Samuel, R. Gupta, D. Chandra, Grid interface of wind power with large split-winding alternator using cascaded multilevel inverter, IEEE Trans. Energy Conv. 26 (1) (2011) 299–309.

[15] M.M. Renge, H.M. Suryawanshi, Five- level diode clamped inverter to eliminate common mode voltage and reduce dv/dt in medium voltage rating induction motor drives, IEEE Trans. Pow. Electron. 23 (4) (2008) 1598–1607.

[16] J. Pereda, J. Dixon, High- frequency link: a solution for using only one DC source in asymmetric cascaded multilevel inverters, IEEE Trans. Industrial Electron. 58 (9) (2011) 3884–3892.

[17] R. Rabinovici, D. Baimel, J. Tomasik, A. Zuckerberger, Series space vector modulation for multi-level cascaded H-bridge inverters, IET Power Electron. 3 (6) (2010) 843–857.

[18] J. Kumar, B. Das, P. Agarwal, Optimized switching scheme of a cascade multi-level Inverter, Electr. Pow. Compo. Sys. 38 (4) (2010) 445–464.

[19] M. Marchesoni, P. Tenca, Diode-clamped multilevel converters: a practicable way to balance DC-link voltages, IEEE Trans. Industrial Electron. 49 (4) (2002) 752–765.

[20] J. Kumar, P. Agarwal, Implementation of cascade multilevel inverter-based STATCOM, IETE J. Res. 56 (2) (2010) 119–128.

[21] J. Pou, R. Pindado, D. Boroyevich, Voltage-balance limits in four-level diode-clamped converters with passive front ends, IEEE Trans. Industrial Electron. 52 (1) (2005) 190–196.

[22] A. Yazdani, R. Iravani, A neutral-point clamped converter system for direct-drive variable-speed wind power unit, IEEE Trans. Energy Conv. 21 (2) (2006) 596–607.

[23] P. Zhiguo, F.Z. Peng, K.A. Corzine, V.R. Stefanovic, J.M. Leuthen, S. Gataric, Voltage balancing control of diode-clamped multilevel rectifier/inverter systems, IEEE Trans. Industrial Electron. 41 (6) (2005) 1698–1706.

[24] B.R. Andersen, L. Xu, P.J. Horton, P. Cartwright, Topologies for VSC transmission, Power Eng. J. 16 (3) (2002) 142–150.

[25] H. Akagi, T. Hatada, Voltage balancing control for a three-level diode-clamped converter in a medium-voltage transformerless hybrid active filter, IEEE Trans. Pow. Electron. 24 (3) (2009) 571–579.

[26] A. Yazdani, R. Iravani, Dynamic model and control of the NPC-based back-to-back HVDC system, IEEE Trans. Pow. Deliv. 21 (1) (2006) 414–424.

[27] P. Chaturvedi, S. Jain, P. Agarwal, Carrier based neutral point potential regulator with reduced switching losses for three-level diode clamped inverter, IEEE Trans. Industrial Electron. 61 (2) (2014) 613–624.

[28] A. Bendre, S. Krstic, J. Vander Meer, G. Venkataramanan, Comparative evaluation of modulation algorithms for neutral-point-clamped converters, IEEE Trans. Industry Appl. 41 (2) (2005) 634–643.

[29] M.M. Renge, H.M. Suryawanshi, Three-dimensional space-vector modulation to reduce common-mode voltage for multilevel inverter, IEEE Trans. Industrial Electron. 57 (7) (2010) 2324–2331.

[30] N. Jayaram, P. Agarwal, S. Das, A three phase five level cascaded H-Bridge rectifier with zero current injection scheme, in: Power Electronics, Drives and Energy Systems (PEDES), 2012 IEEE International Conference on, pp. 1–7, 2012.

[31] K. Xiaomin, K.A. Corzine, Y.L. Familiant, A unique fault-tolerant design for flying capacitor multilevel inverter, IEEE Trans. Pow. Electron. 19 (4) (2004) 979–987.

[32] C. Feng, J. Liang, V.G. Agelidis, Modified phase-shifted PWM control for flying capacitor multilevel converters, IEEE Trans. Pow. Electron. 22 (1) (2007) 178–185.

[33] P. Lezana, R. Aguilera, D.E. Quevedo, Model predictive control of an asymmetric flying capacitor converter, IEEE Trans. Industrial Electron. 56 (6) (2009) 1839–1846.

[34] A. Shukla, A. Ghosh, A. Joshi, Improved multilevel hysteresis current regulation and capacitor voltage balancing schemes for flying capacitor multilevel inverter, IEEE Trans. Pow. Electron. 23 (2) (2008) 518–529.

[35] J. Huang, K.A. Corzine, Extended operation of flying capacitor multilevel inverters, IEEE Trans. Pow. Electron. 21 (1) (2006) 140–147.
[36] M. Khazraei, H. Sepahvand, K.A. Corzine, M. Ferdowsi, Active capacitor voltage balancing in single-phase flying-capacitor multilevel power converters, IEEE Trans. Industrial Electron. 59 (2) (2012) 769–778.
[37] V.G. Agelidis, D.M. Baker, W.B. Lawrance, C.V. Nayar, A multilevel PWM inverter topology for photovoltaic applications, in: Industrial Electronics, 1997. ISIE '97. Proceedings of the IEEE International Symposium on, vol.2, no., pp. 589–594, July 7–11, 1997.
[38] M.D. Manjrekar, T.A. Lipo, A hybrid multilevel inverter topology for drive applications, in: Applied Power Electronics Conference and Exposition, 1998. Thirteenth Annual, vol. 2, no., pp. 523–529, February 15–19, 1998.
[39] Y. Xiaoming, I. Barbi, Fundamentals of a new diode clamping multilevel inverter, IEEE Trans. Pow. Electron. 15 (4) (2000) 711–718.
[40] F.Z. Peng, A generalized multilevel inverter topology with self voltage balancing, IEEE Trans. Industry Appl. 37 (2) (2001) 611–618.
[41] Y. Cheng, M.L. Crow, A diode-clamped multi-level inverter for the StatCom/BESS, in: Power Engineering Society Winter Meeting, 2002. IEEE, vol. 1, pp. 470–475, 2002.
[42] S. Mariethoz, A. Rufer, New configurations for the three-phase asymmetrical multilevel inverter, in: Industry Applications Conference, 2004. 39th IAS Annual Meeting. Conference Record of the 2004 IEEE, vol. 2, pp. 828–835, October 3–7, 2004.
[43] M. Veenstra, A. Rufer, Control of a hybrid asymmetric multilevel inverter for competitive medium-voltage industrial drives, IEEE Trans. Industry Appl. 41 (2) (March–April 2005) 655–664.
[44] A. Chen, L. Hu, L. Chen, Y. Deng, X. He, A multilevel converter topology with fault-tolerant ability, IEEE Trans. Pow. Electron. 20 (2) (2005) 405–415.
[45] A. Chen, X. He, Research on hybrid-clamped multilevel-inverter topologies, IEEE Trans. Industrial Electron. 53 (6) (2006) 1898–1907.
[46] M.S.W. Chan, K.T. Chau, A new switched-capacitor boost-multilevel inverter using partial charging, IEEE Trans. Circuits Syst II: Express Briefs 54 (12) (2007) 1145–1149.
[47] G. Grandi, D. Ostojic, C. Rossi, A. Lega, Control strategy for a multilevel inverter in grid-connected photovoltaic applications, in: Electrical Machines and Power Electronics, 2007. ACEMP '07. International Aegean Conference on, pp. 156–161, September 10–12, 2007.
[48] P. Lezana, J. Rodriguez, D.A. Oyarzun, Cascaded multilevel inverter with regeneration capability and reduced number of switches, IEEE Trans. Industrial Electron. 55 (3) (2008) 1059–1066.
[49] S. Daher, J. Schmid, F.L.M. Antunes, Multilevel inverter topologies for stand-alone PV systems, IEEE Trans. Industrial Electron. 55 (7) (2008) 2703–2712.
[50] R. Gonzalez, E. Gubia, J. Lopez, L. Marroyo, Transformerless single-phase multilevel-based photovoltaic inverter, IEEE Trans. Industrial Electron. 55 (7) (2008) 2694–2702.
[51] K. Wang, Y. Li, Z. Zheng, A new transformerless cascaded multilevel converter topology, in: Energy Conversion Congress and Exposition, 2009. ECCE 2009. IEEE, pp. 3124–3129, September 20–24, 2009.
[52] Z. Du, B. Ozpineci, L.M. Tolbert, J.N. Chiasson, DC–AC cascaded H- bridge multilevel boost inverter with no inductors for electric/hybrid electric vehicle applications, IEEE Trans. Industry Appl. 45 (3) (2009) 963–970.

[53] J. Ewanchuk, J. Salmon, A.M. Knight, Performance of a high-speed motor drive system using a novel multilevel inverter topology, IEEE Trans. Industry Appl. 45 (5) (2009) 1706–1714.

[54] Y. Liu, F.L. Luo, Multilevel inverter with the ability of self-voltage balancing, Electr. Pow. Appl. IEE Proc. 153 (1) (2006) 105–115.

[55] S. De, D. Banerjee, K. Siva Kumar, K. Gopakumar, R. Ramchand, C. Patel, Multilevel inverters for low-power application, IET Pow. Electr. 4 (4) (2011) 384–392.

[56] C.-C. Hua, C.-W. Wu, C.-W. Chuang, A novel DC voltage charge balance control for cascaded inverters, IET Pow. Electr. 2 (2) (2009) 147–155.

[57] A.R. Beig, A. Dekka, Experimental verification of multilevel inverter-based standalone power supply for low-voltage and low-power applications, IET Pow. Electr. 5 (6) (2012) 635–643.

[58] S. Gui-Jia, Multilevel DC-link inverter, IEEE Trans. Industry Appl. 41 (3) (2005) 848–854.

[59] S. Gui-Jia, Multilevel DC link inverter, in: Industry Applications Conference, 2004. 39th IAS Annual Meeting. Conference Record of the 2004 IEEE, vol. 2, pp. 806–812, October 3–7, 2004

[60] G. Ceglia, V. Guzman, C. Sanchez, F. Ibanez, J. Walter, M.I. Gimenez, A new simplified multilevel inverter topology for DC-AC conversion, IEEE Trans. Pow. Electron. 21 (5) (2006) 1311–1319.

[61] N.A. Rahim, K. Chaniago, J. Selvaraj, Single-phase seven-level grid-connected inverter for photovoltaic system, IEEE Trans. Industrial Electron. 58 (6) (2011) 2435–2444.

[62] G.M. Martins, J.A. Pomilio, S. Buso, G. Spiazzi, Three-phase low-frequency commutation inverter for renewable energy systems, IEEE Trans. Industrial Electron. 53 (5) (2006) 1522–1528.

[63] Y. Hinago, H. Koizumi, A single phase multilevel inverter using switched series/parallel DC voltage sources, in: Energy Conversion Congress and Exposition, 2009. ECCE 2009. IEEE, pp. 1962–1967, September 20–24, 2009.

[64] Y. Hinago, H. Koizumi, A single-phase multilevel inverter using switched series/parallel DC voltage sources, IEEE Trans. Industrial Electron. 57 (8) (2010) 2643–2650.

[65] W. Choi, F. Kang, H-bridge based multilevel inverter using PWM switching function, in: Telecommunications Energy Conference, 2009. INTELEC 2009. 31st International, pp. 1–5, October 18–22, 2009.

[66] S.H. Lee, F.S. Kang, A new structure of H-bridge multilevel inverter, in: Proceedings of the KIPE Conference, 2008, pp. 388–390, 2008.

[67] E. Babaei, A cascade multilevel converter topology with reduced number of switches, IEEE Trans. Pow. Electron. 23 (6) (2008) 2657–2664.

[68] Y. Ounejjar, K. Al-Haddad, A novel high energetic efficiency multilevel topology with reduced impact on supply network, in: Industrial Electronics, 2008. IECON 2008. 34th Annual Conference of IEEE, pp. 489–494, November 10–13, 2008.

[69] Y. Ounejjar, K. Al-Haddad, A new high power efficiency cascaded U cells multilevel converter, in: Industrial Electronics, 2009. ISIE 2009. IEEE International Symposium on, pp. 483–488, July 5–8, 2009.

[70] Y. Ounejjar, K. Al-Haddad, L.A. Gregoire, Novel three phase seven level PWM converter, in: Electrical Power and Energy Conference (EPEC), 2009 IEEE, pp. 1–6, 22–23, 2009.

[71] Y.Ounejjar, K. Al-Haddad, Multilevel hysteresis controller of the novel seven-level packed U cells converter, in: Power Electronics Electrical Drives Automation and Motion (SPEEDAM), 2010 International Symposium on, pp. 186–191, June 14–16, 2010.

[72] Y. Ounejjar, K. Al-Haddad, L.A. Grégoire, Packed U cells multilevel converter topology: theoretical study and experimental validation, IEEE Trans. Industrial Electron. 58 (4) (2011) 1294–1306.

[73] J. Ebrahimi, E. Babaei, G.B. Gharehpetian, A new multilevel converter topology with reduced number of power electronic components, IEEE Trans. Industrial Electron. 59 (2) (2012) 655–667.

[74] E. Najafi, A. Yatim, Design and implementation of a new multilevel inverter topology, IEEE Trans. Industrial Electron. 59 (11) (2012) 4148–4154.

[75] E. Najafi, A. Yatim, A.S. Samosir, A new topology—reversing voltage (RV)—for multi level inverters, in: Power and Energy Conference, 2008. PECon 2008. IEEE 2nd International, pp. 604–608, December 1–3, 2008.

[76] M.F. Kangarlu, E. Babaei, A generalized cascaded multilevel inverter using series connection of submultilevel inverters, IEEE Trans. Pow. Electron. 28 (2) (2013) 625–636.

Chapter 4

Universal Control Scheme with Voltage-Level-Based Methods

4.1 INTRODUCTION

Multilevel inverters (MLIs) are circuits which convert multiple input-DC levels into a stepped AC waveform. For this, power electronic switches function as controlled components. These switches are operated only in two modes—either ON (conducting) or OFF (blocking). A valid switching combination in a given power electronic converter is called a *state*. A given converter can have numerous valid states leading to respective voltages at the output terminals. *The process in which the converter undergoes systematic transitions in states with respect to time is called modulation.* A good number of modulation strategies have been the subject of intense research for many years for the control of MLIs [1–13]. For any given topology, there are preferred modulation strategies associated with it [14–17]. Apart from the classical topologies, many new topologies have also evolved [18–28]. While reading a literature dealing with a specific MLI topology, you will often notice that the author describes a modulation that is very specific to that topology. Such a switching procedure may not directly applicable to other topologies.

The primary objective of this chapter is to formulate a common modulation scheme which can be used to study *all* topologies. Utilizing this scheme for secondary (but important) features like charge balance control, optimal switching, utilization of specified states, and so on are demonstrated in subsequent chapters.

In this chapter, a brief introduction to popular "voltage-level-based" modulation schemes for MLIs is first presented. Then the development of a generalized control scheme is presented which is applicable to all topologies and is therefore referred to as a "universal control scheme (UCS) using voltage-level-based methods." With the help of simulation and experimental studies, the applicability of this scheme is ascertained.

Multilevel Inverters. DOI: https://doi.org/10.1016/B978-0-12-812448-2.00004-0

4.2 MODULATION STRATEGIES FOR MLIS

Several modulation methods have been proposed in the literature for MLIs. A categorization of these is presented in Reference [29] and updated in Reference [30]. A basic categorization of these schemes is shown in Fig. 4.1.

These methods can also be categorized based on switching frequency as:

1. Low switching frequency-based methods (space vector control, selective harmonic elimination (SHE), and nearest level control);

2. High switching frequency-based methods (space vector pulse-width modulation [SVPWM] and multicarrier PWM).

Since the space vector-based modulation methods become cumbersome for greater than five levels, this method has not been discussed in this book. In this section, multicarrier PWM is discussed along with the "SHE" as a special case of carrier-based modulation schemes.

Multilevel carrier-based PWM methods have multiple carriers which can be triangular or saw-tooth signals. Carrier signals present freedom in the following parameters: frequency, amplitude, phase of each carrier, and offset between the carriers. The reference wave (modulating signal) also presents

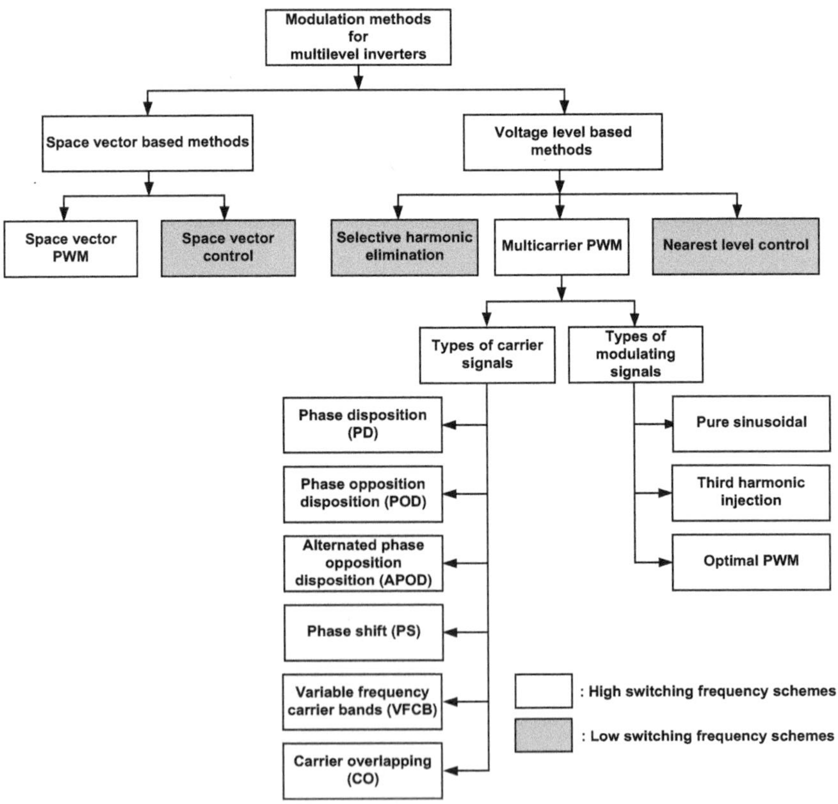

FIGURE 4.1 Categorization of modulation strategies for MLIs.

freedom in frequency, amplitude, phase angle, and injection of zero sequence signals to it. Thus, by using these combinations, many multilevel carrier-based PWM methods can be obtained. In general, an inverter with n_{levels} number of levels (including zero level) in the phase voltage, would require a set of $n_{levels} - 1$ number of carrier signals. Multicarrier PWM methods can be classified according to carrier signals and modulating signal as described in the forthcoming subsections.

4.2.1 Multicarrier PWM with Different Carrier Signals

Various multicarrier PWM techniques can be derived based on the placement of carrier signals, their magnitudes and frequency, and their overlapping duration with each other. Some of these methods are:

1. Phase Disposition PWM

In this technique, all carrier signals are in phase and level-shifted. The zero reference is placed at the middle of the carrier set. For pulse generation for a five-level inverter, the set of phase-disposed carrier waveforms is shown in Fig. 4.2.

2. Phase Opposition Disposition PWM

In this scheme, the carrier signals above the zero reference are in the same phase and the carrier signals below the zero reference are in the same phase but the carriers below and above the zero reference are phase-shifted by 180 degrees, as shown in Fig. 4.3 for a five-level inverter.

3. Alternate Phase Opposition Disposition PWM (APOD-PWM)

As the name suggests, the carrier signals are so disposed that they are alternately phase-displaced from each other by 180 degrees, as illustrated in Fig. 4.4 for a five-level inverter.

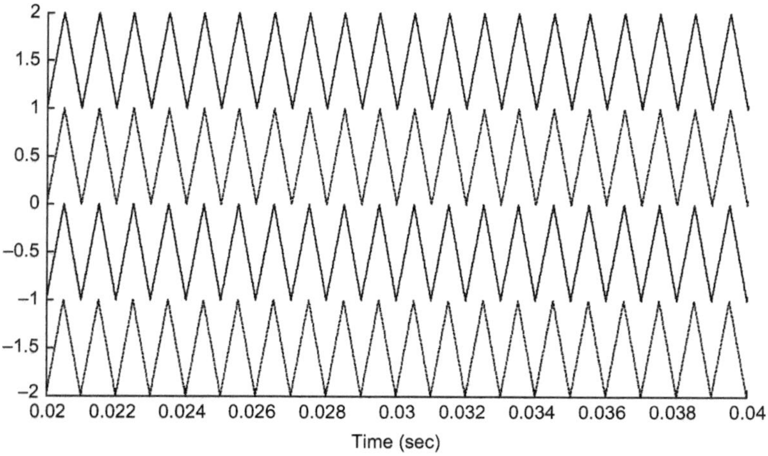

FIGURE 4.2 Carrier signals in a phase-disposition scheme.

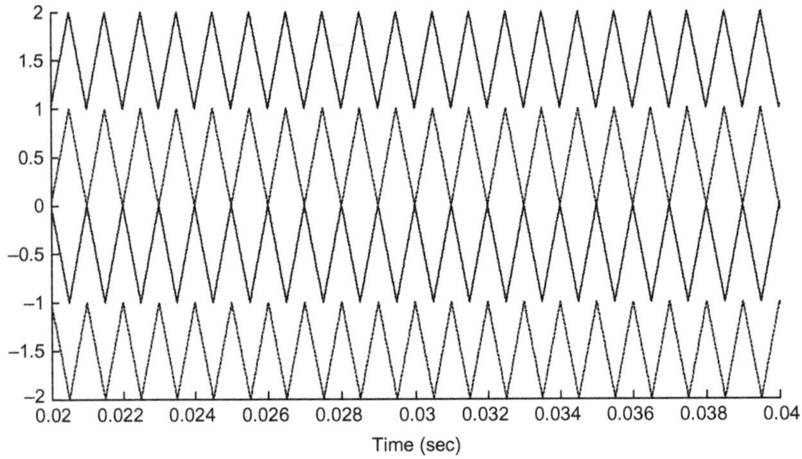

FIGURE 4.3 Carrier signals in a phase-opposition-disposition (POD) scheme.

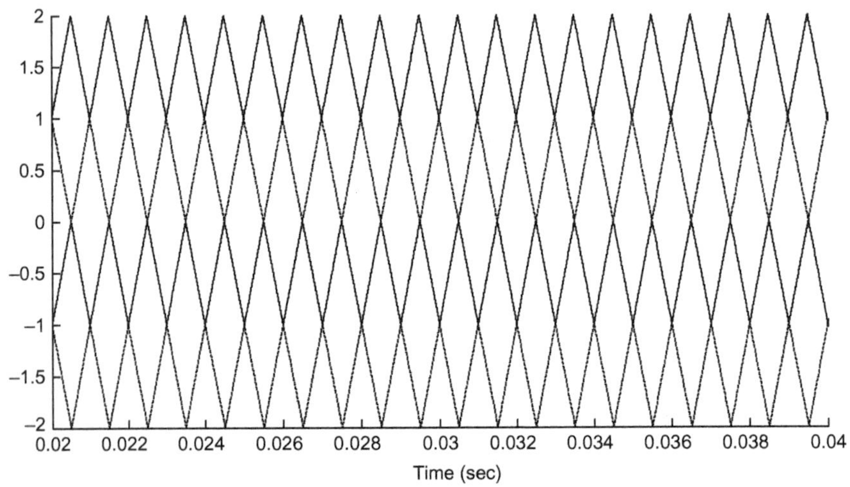

FIGURE 4.4 Carrier signals in an alternate phase-opposition-disposition (APOD) scheme.

4. Phase Shift PWM (PS-PWM)

In this scheme, all the carriers are phase-shifted at an appropriate angle as shown in Fig. 4.5. The performance with this technique depends on the phase shift angle between carriers.

5. Variable Frequency Carrier Bands PWM (VFCB-PWM)

In this technique, the frequency of all the carriers is not same. Some carriers have different frequencies to others as shown in Fig. 4.6. The techniques mentioned above in (1)–(4) can be modified to achieve VFCB-PWM.

6. Carrier Overlapping PWM

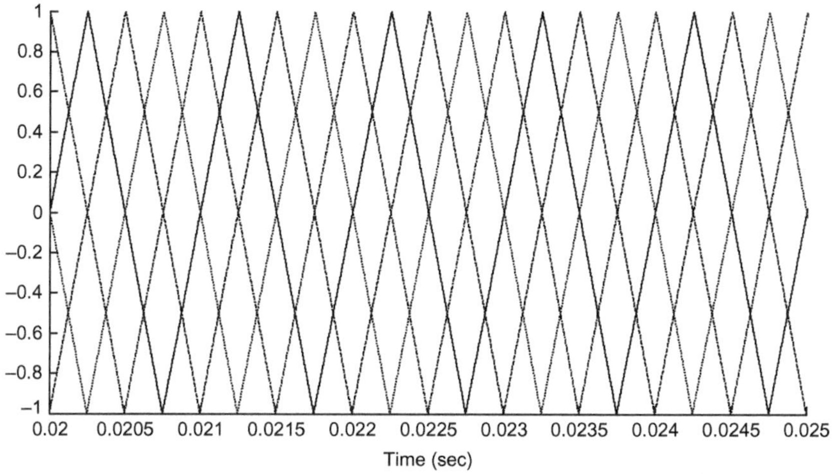

FIGURE 4.5 Carrier signals in a phase shift (PS) scheme.

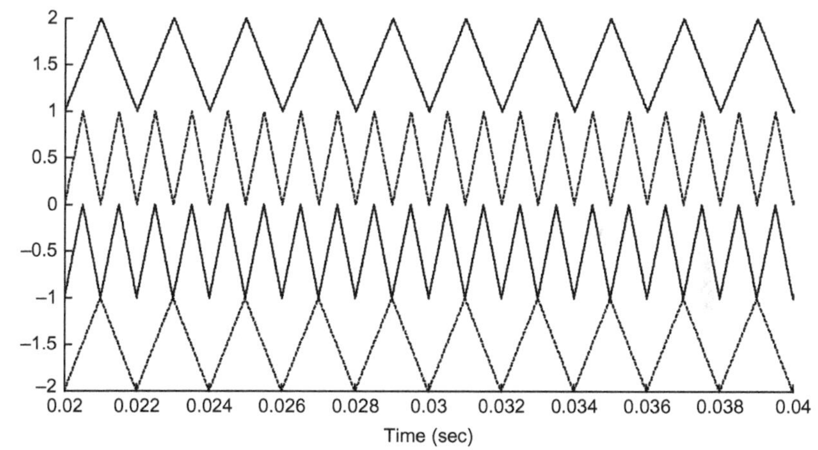

FIGURE 4.6 Carrier signals in a variable frequency carrier bands (VFCB) scheme.

In this scheme, the carriers overlap each other as illustrated in Fig. 4.7. The amount of overlapping decides the quality of the output waveform.

4.2.2 Multicarrier PWM with Different Modulating Signals

Another classification of multicarrier PWM techniques for MLIs is based on the nature of the modulating signal. Purely sinusoidal or appropriately preprocessed modulating signals that can be utilized for switching are discussed here.

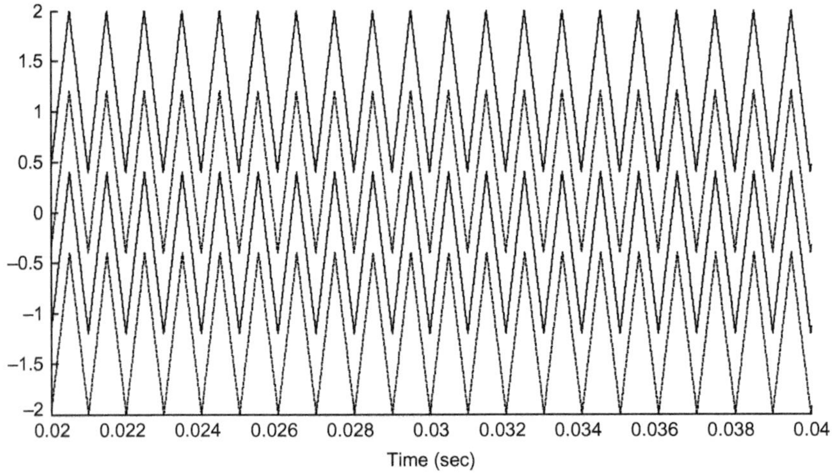

FIGURE 4.7 Carrier signals in a carrier overlapping (CO) scheme.

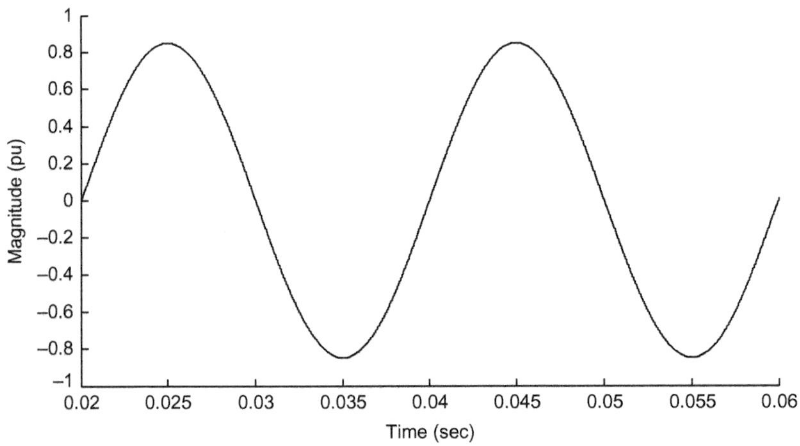

FIGURE 4.8 A pure sinusoidal modulating signal.

1. Pure Sinusoidal PWM

This is the most widely accepted modulating signal used with triangular carriers to generate control pulses. A sinusoidal modulating signal is shown in Fig. 4.8.

2. Third Harmonic Injection PWM

This method is used in three-phase systems where the effect of the injection of third harmonic is nullified in the output voltage spectrum. By adding the third harmonic component to the fundamental component, the gain of the modulator can be increased. The amount of gain increase depends on the magnitude of the third harmonic content. It is observed that injecting 17% of

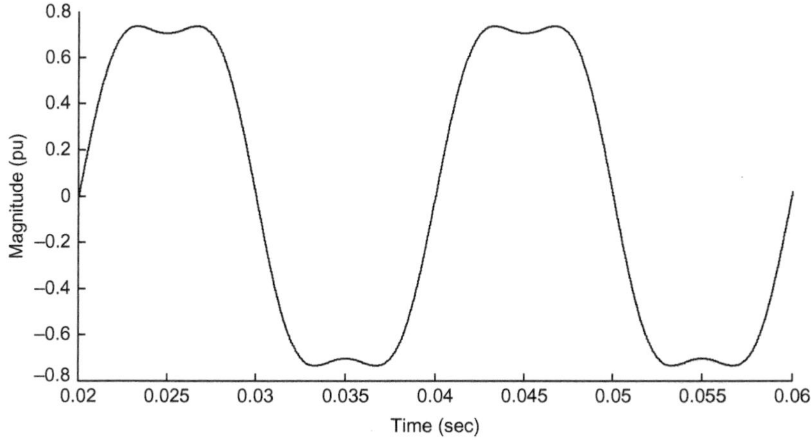

FIGURE 4.9 Modulating signal with a third harmonic injection sinusoidal PWM technique.

the third harmonic into the pure sinusoidal wave yields approximately a 15% increase in gain over the pure sinusoidal reference. A modulating signal with third harmonic injection is illustrated in Fig. 4.9.

3. Optimal PWM

The optimal PWM technique is used to reduce the switching losses and to optimize the harmonic contents in inverter voltage. The basic principle of this technique is to add some variable offset signal to the modulating wave to generate modified control pulses, thereby changing the harmonic profile of the output voltage. This offset signal may be [31],

$$v_{\text{offset}} = -\frac{[\min(v_r, v_y, v_b) + \max(v_r, v_y, v_b)]}{2} \tag{4.1}$$

where, v_r, v_y, and v_b are the sinusoidal modulating signals of the R, Y, and B phases, respectively, of a three-phase system. The new modulating signals are obtained as:

$$\begin{aligned} v_{r,\text{SFO}} &= v_r - v_{\text{offset}}; \\ v_{y,\text{SFO}} &= v_y - v_{\text{offset}}; \text{ and} \\ v_{b,\text{SFO}} &= v_b - v_{\text{offset}} \end{aligned} \tag{4.2}$$

"Preprocessing" as described above is illustrated in Fig. 4.10A and the modulating signal with optimal sinusoidal PWM technique is shown in Fig. 4.10B.

4.2.3 SHE Technique

SHE is a low switching frequency technique in which one or two commutations of the power semiconductor switches are performed during one cycle of the output voltages [32–33], generating a staircase waveform, as illustrated in Fig. 4.11A.

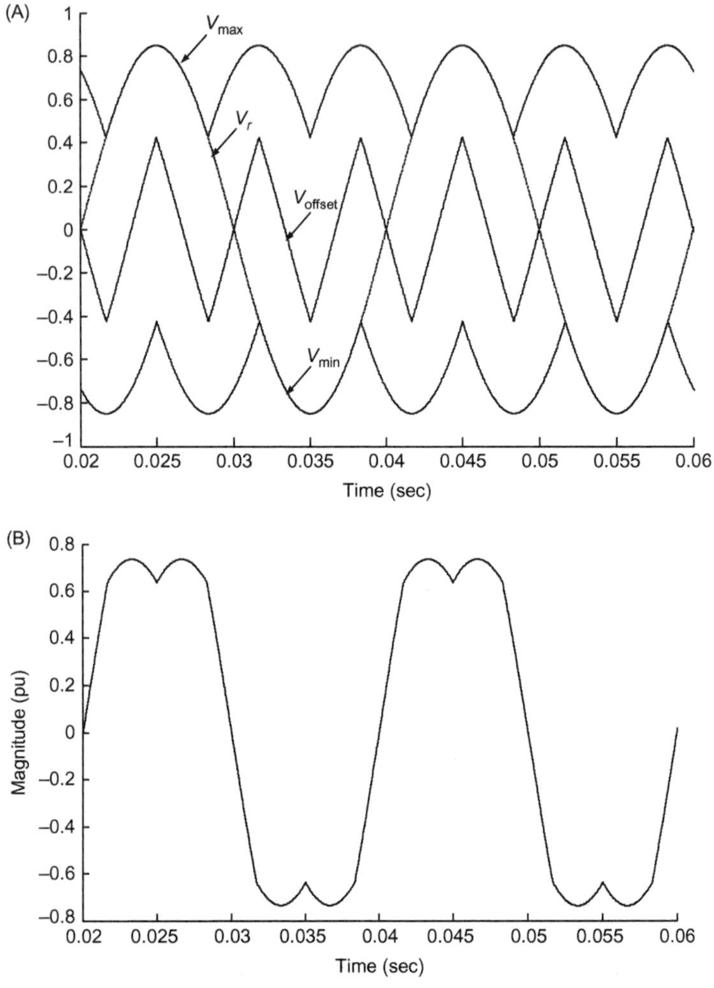

FIGURE 4.10 (A) Signals v_{max}, v_{min}, v_r, and off-set signal v_{offset} in a PD SFO-PWM scheme; (B) modulating signal for an SFO-PWM scheme.

For example, a quarter-wave symmetric stepped voltage waveform synthesized by a nine-level inverter needs to be programmed for four switching angles, namely α_1, α_2, α_3, and α_4, as indicated in Fig. 4.11A. By applying Fourier series analysis, the amplitude of any odd harmonic of the stepped waveform can be expressed as:

$$h_n = \frac{4V_{dc}}{n\pi}\left[\sum_{k=1}^{4}\cos n\alpha_k\right] \tag{4.3}$$

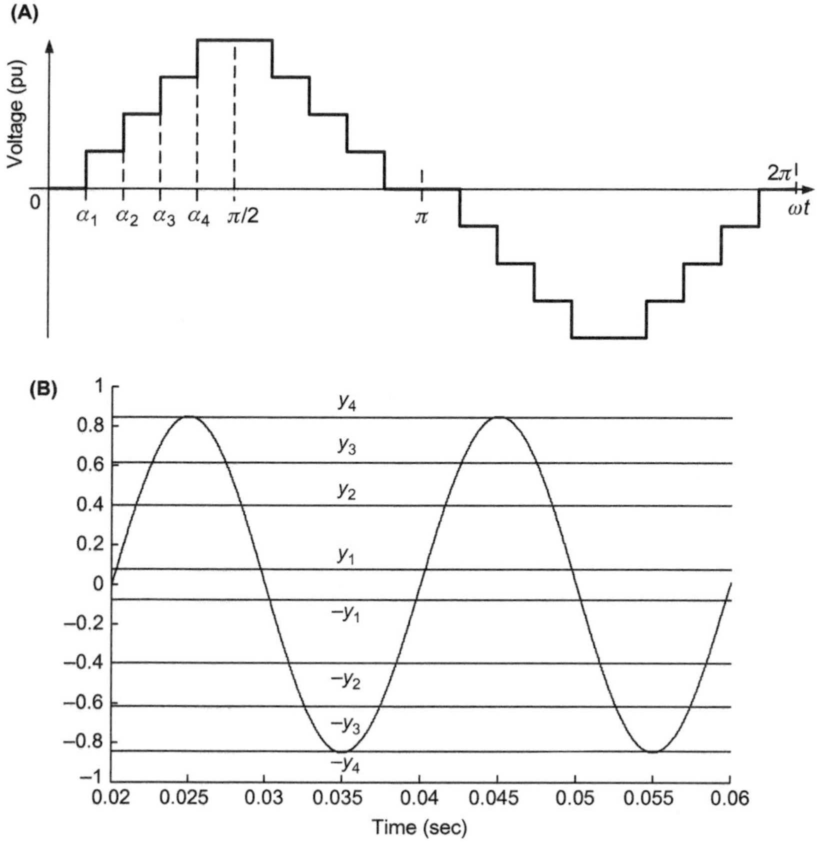

FIGURE 4.11 (A) Nine-level multistepped waveform with quarter wave symmetry. (B) Modulating and carrier signals representing the SHE method.

where the amplitudes of all even harmonics are zero and h_n is an odd harmonic order. The switching angles must satisfy the condition:

$$\alpha_1 < \alpha_2 < \alpha_3 < \alpha_4 < \pi/2 \tag{4.4}$$

To minimize the harmonic distortion and to achieve adjustable amplitude of the fundamental component, selected harmonic contents can be removed from the voltage waveform.

Assuming a modulation index of 0.85 and elimination of third, fifth, and seventh harmonics, the following equations are obtained:

$$\begin{aligned}
&\cos\alpha_1 + \cos\alpha_2 + \cos\alpha_3 + \cos\alpha_4 = (0.85)3\pi/4;\\
&\cos 3\alpha_1 + \cos 3\alpha_2 + \cos 3\alpha_3 + \cos 3\alpha_4 = 0;\\
&\cos 5\alpha_1 + \cos 5\alpha_2 + \cos 5\alpha_3 + \cos 5\alpha_4 = 0;\ \text{and}\\
&\cos 7\alpha_1 + \cos 7\alpha_2 + \cos 7\alpha_3 + \cos 7\alpha_4 = 0
\end{aligned} \tag{4.5}$$

Solving these equations gives following solution:

$$\alpha_1 = 5.25^\circ,\ \ \alpha_2 = 28.12^\circ,\ \ \alpha_3 = 46.38^\circ \text{ and } \alpha_4 = 84.09^\circ \qquad (4.6)$$

SHE technique can be thought of as a special case of multicarrier modulation where carriers are DC signals with specific magnitudes and a sinusoidal reference. For the case of a nine-level waveform as discussed above, reference and carrier waveforms $\{y_{1,2,3,4} = 0.85\sin(\alpha_{1,2,3,4})\}$ corresponding to the aforementioned values are shown in Fig. 4.11B.

4.3 DESCRIPTION OF THE UCS

In this section, the so-called UCS is explained. It is developed in a manner that it can be applied to all topologies of MLIs. A schematic diagram of a UCS is shown in Fig. 4.12.

For a given inverter, let the number of levels in the phase voltage be n_{levels}. A voltage source inverter allows multilevel operation if $n_{\text{levels}} \geq 3$. Since level 0 is obviously important, n_{levels} is considered to be odd. Also, the number of positive levels in an n_{levels} waveform will be:

$$n = (n_{\text{levels}} - 1)/2 \qquad (4.7)$$

As discussed in the previous section, carrier and modulating signals must be appropriately chosen. The modulating (or reference) signal, $f_{\text{ref}}\,(t)$,

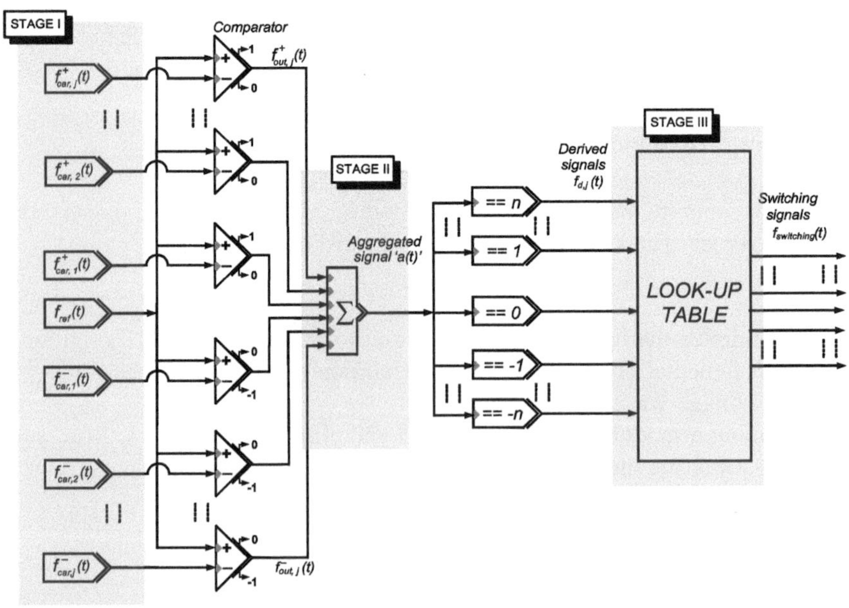

FIGURE 4.12 Universal control scheme.

is a waveform with amplitude A_{ref} and angular frequency ω_{ref}. There should be $2n$ carrier signals for n_{levels} number of output levels. These carriers have the angular frequency ω_{car} and peak-to-peak amplitude A_{car}. Carrier signals above the zero reference are designated as $f^{+}_{car,j}\,(t)$ and those below the zero reference are designated as $f^{-}_{car,j}(t)$, $\{ j=1$ to $n \}$. Carrier signals are disposed such that they occupy contiguous bands and the zero reference is placed in the middle.

Accordingly the following quantities can be defined:

$$\text{Frequency modulation index}, m_f = \omega_{car}/\omega_{ref} \tag{4.8}$$

where "m_f" dictates the switching frequency of the power switches and the harmonic profile of the output waveform.

$$\text{Amplitude modulation index}, m_a = A_{ref}/(nA_{car}) \tag{4.9}$$

where "m_a" dictates the number of levels in the output waveform and its peak value.

At every instant, each carrier is compared with the modulating signal. For all the carrier signals above the zero reference, each comparator gives "1" if the modulating signal is greater than the carrier and "0" otherwise. For all the carrier signals below the zero reference, each comparison gives "0" if the modulating signal is greater than the carrier and " − 1" otherwise. That is,

$$\begin{aligned} f^{+}_{out,j}(t) &= 1,\ \text{for}\, f_{ref}(t) \geq f^{+}_{car,j}(t) \\ &= 0,\ \text{otherwise} \end{aligned} \tag{4.10}$$

$$\begin{aligned} f^{-}_{out,j}(t) &= 0,\ \text{for}\, f_{ref}(t) \geq f^{-}_{car,\,j}(t) \\ &= -1,\ \text{otherwise} \end{aligned} \tag{4.11}$$

The results so obtained are added, thereby synthesizing a signal called "aggregated signal" designated as $a(t)$. That is,

$$a(t) = \sum_{j=1}^{n} \{ f^{+}_{out,\,j}(t) + f^{-}_{out,\,j}(t) \} \tag{4.12}$$

It is extremely important to note here that $a(t)$ acquires the same waveshape as that of the expected output voltage.

The actual driving signals for the power electronic switches in a given topology need to be derived from $a(t)$ using logic circuits and a look-up table. To do this, Boolean functions, $f_{d,\,j}(t)$, are derived from $a(t)$ using the following criteria:

$$\begin{aligned} f_{d,\,j}(t) &= 1,\ \text{if}\ a(t) = = j \\ &= 0,\ \text{otherwise; where,}\ j = -n \text{ to } +n \end{aligned} \tag{4.13}$$

Thus, a total of n_{levels} derived signals will be obtained and each is used to drive those switches which have to remain ON for the corresponding voltage level in the output waveform. Consequently, the switching function $f_{switching}(t)$, for a given switch, is obtained by feeding appropriate derived signals into an OR gate. Mathematically, the switching function can be expressed as:

$$f_{switching}(t) = \overline{\prod \overline{f_d(t)}}, \tag{4.14}$$

where,

$$\overline{f_d(t)} = 1 - f_d(t) \tag{4.15}$$

For the APOD arrangement of triangular carrier signals and a pure sinusoidal modulating signal for a five-level inverter, waveforms at various stages of the scheme are illustrated in Figs. 4.13 and 4.14. The overall scheme is marked with three stages as shown in Fig. 4.12. The three stages are:

1. *Stage I*: Appropriate selection of carriers and reference enables the UCS to incorporate any modulation strategy discussed in Section 4.2. This stage can also be modified to alter the aggregated signal *a(t)* so that state utilization of a given topology can be strategized, as will be illustrated in later chapters.
2. *Stage II*: Appropriate manipulations with the aggregated signal "*a(t)*" give flexibility to implement charge balance control in input DC sources. An example of such manipulation is demonstrated in Chapters 5, Multilevel Inverter Based on Bridge-Type Connected Sources, and 6, Cross-Connected Sources–Based Multilevel Inverter.
3. *Stage III*: A look-up table can be formulated based on the topology and hence the scheme is applicable to all multilevel topologies. Moreover, depending on the signal *a(t)* and the look-up table, one or more of the redundant states can be chosen for targeted operation of the inverter.

No mention of "nearest level control" has been made yet. It is a low switching frequency scheme in which the output voltage level nearest to the reference is selected and does not use any carrier signals [34]. In the context of a UCS, the methodology inherently synthesizes the aggregated signal "*a (t)*," as the reference signal is multiplied by the number of positive levels and then "round" function is applied, that is:

$$a(t) = \text{round}[n\ m_a\ sin(\omega t)] \tag{4.16}$$

In order to verify the efficacy of UCS, an implementation of the scheme is shown for three different topologies. Although all modulation techniques can be employed for all topologies, application is exemplified for one modulation technique for each topology.

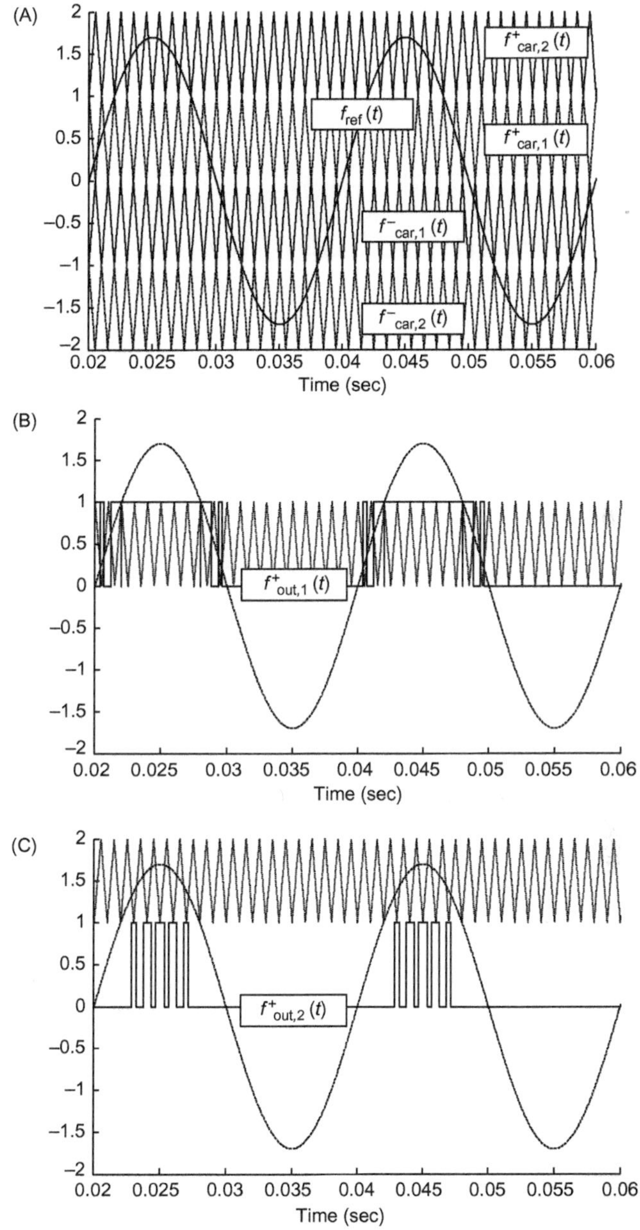

FIGURE 4.13 Waveforms pertaining to UCS for a five-level inverter: (A) reference and carrier waveforms; (B–E) are output signals from comparators; and (F) aggregated signal *a(t)*.

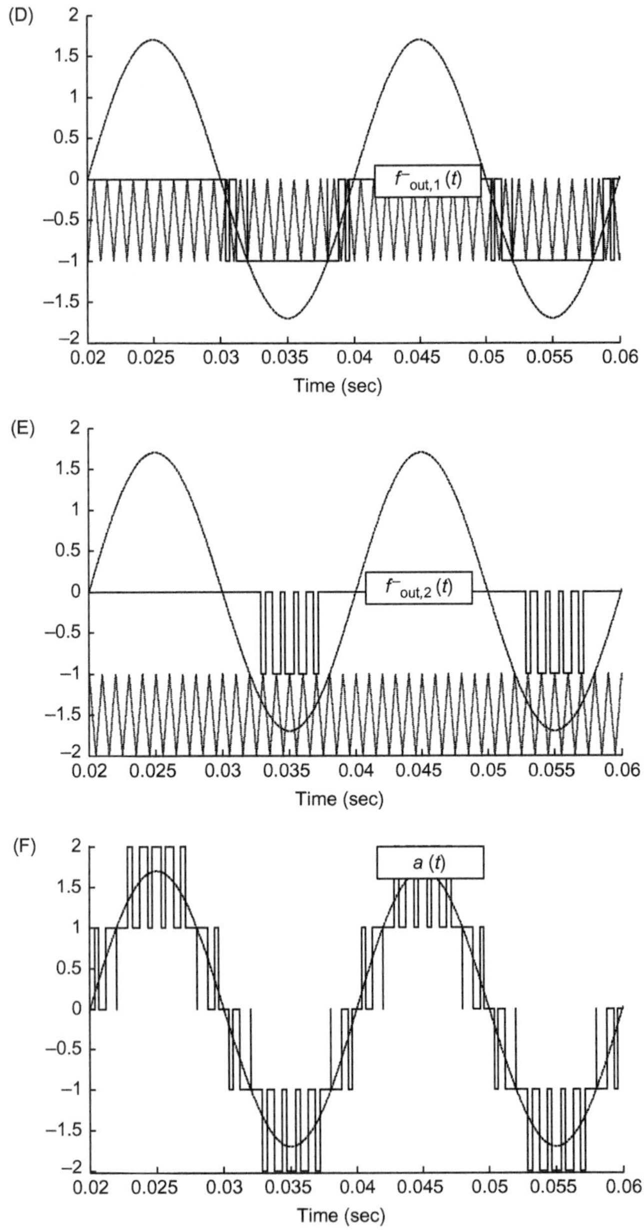

FIGURE 4.13 (Continued)

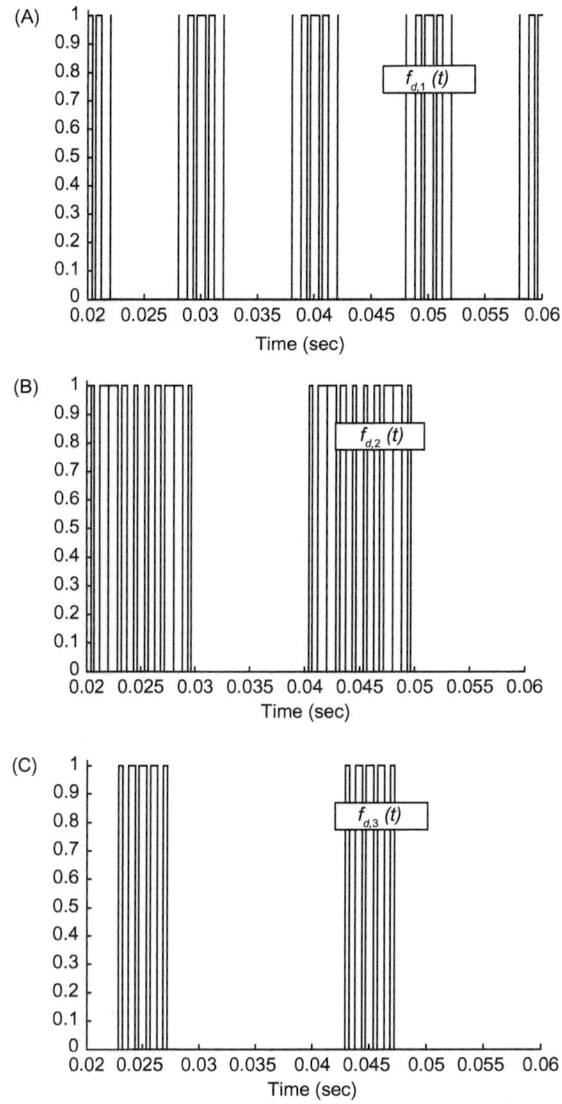

FIGURE 4.14 "Derived signals" as obtained from the aggregated signal *a(t)* for a five-level inverter.

4.4 IMPLEMENTATION OF UCS FOR A FIVE-LEVEL CASCADED H-BRIDGE INVERTER

An example of complete implementation of UCS is demonstrated in this section with the help of a computer simulation and hardware implementation of a five-level cascaded H-bridge inverter.

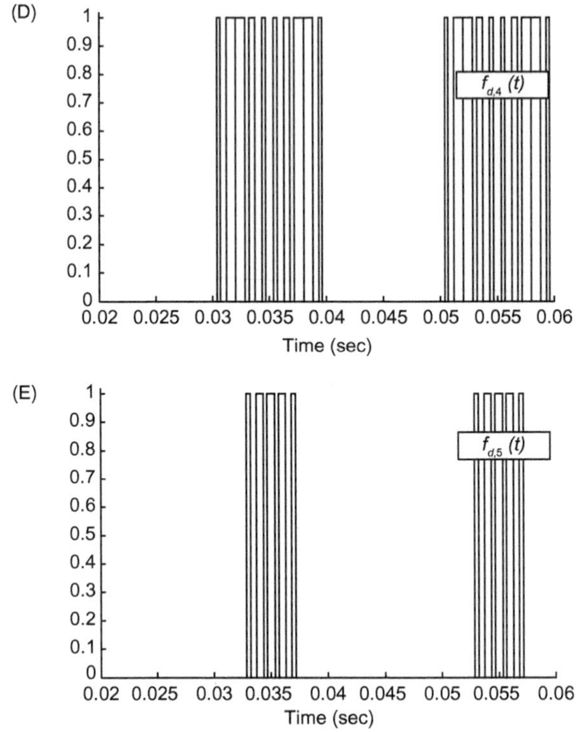

FIGURE 4.14 (Continued)

MATLAB/Simulink along with other toolboxes (like SimPowerSystems) has been used for computer-based modeling. Moreover, MATLAB/Simulink-based models are interfaced with a dSPACE DS1103 real-time controller using Real Time Workshop to generate control pulses for experimental prototypes. Any other equivalent tools, of course, are equally qualified for the purpose.

4.4.1 Simulation Model for Obtaining Aggregated Signal "*a(t)*"

A MATLAB/Simulink model for obtaining aggregated signal *a(t)* for a five-level inverter is shown in Fig. 4.15. Here, a pure sinusoidal reference is used along with four triangular carrier signals disposed in alternate phase opposition disposition such that $m_a = 0.85$ and $m_f = 20$. Waveforms pertaining to the scheme until the synthesis of aggregated signal *a(t)* are illustrated in Fig. 4.13. It can be noted that the signal *a(t)* acquires the wave shape of the output voltage expected at the inverter terminals. Actual driving signals are derived from this signal based on the topology as shown in Section 4.4.2.

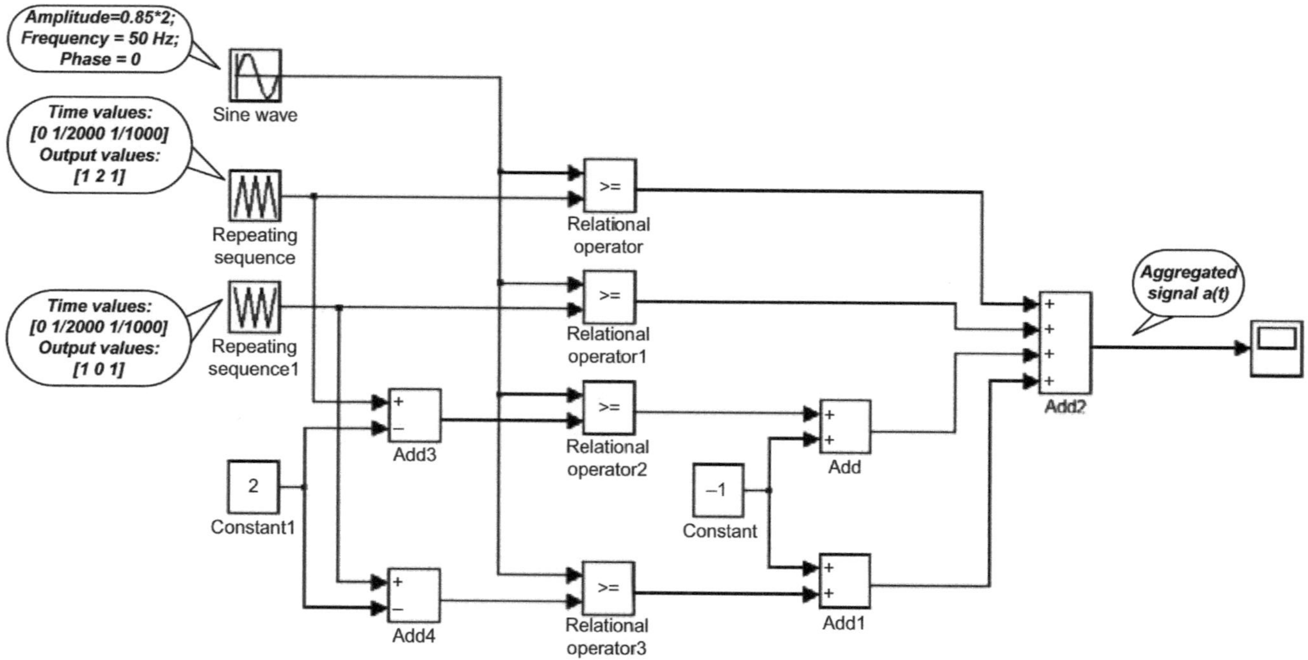

FIGURE 4.15 MATLAB/Simulink model for obtaining aggregated signal $a(t)$ in a five-level sine-PWM inverter.

4.4.2 Simulation Model for Obtaining Actual Driving Pulses for a Five-Level Cascaded H-Bridge Inverter

Actual driving signals are obtained for the aggregated signal *a(t)* using the procedure described in Section 4.3. A MATLAB/Simulink model corresponding to it is shown in Fig. 4.16. First, so-called *derived signals* are obtained from the signal *a(t)*. The number of derived signals is the same as that of the number of levels in the output waveform. Second, actual switching functions depend on the desired association of the output voltage levels with the aggregated signal *a(t)* and the switches which need to simultaneously conducted to obtain these levels. For example, a cascaded H-bridge inverter is shown in Fig. 4.17 with two H-bridges connected in series, each fed with a voltage source V_{DC}. Five levels are expected in the output

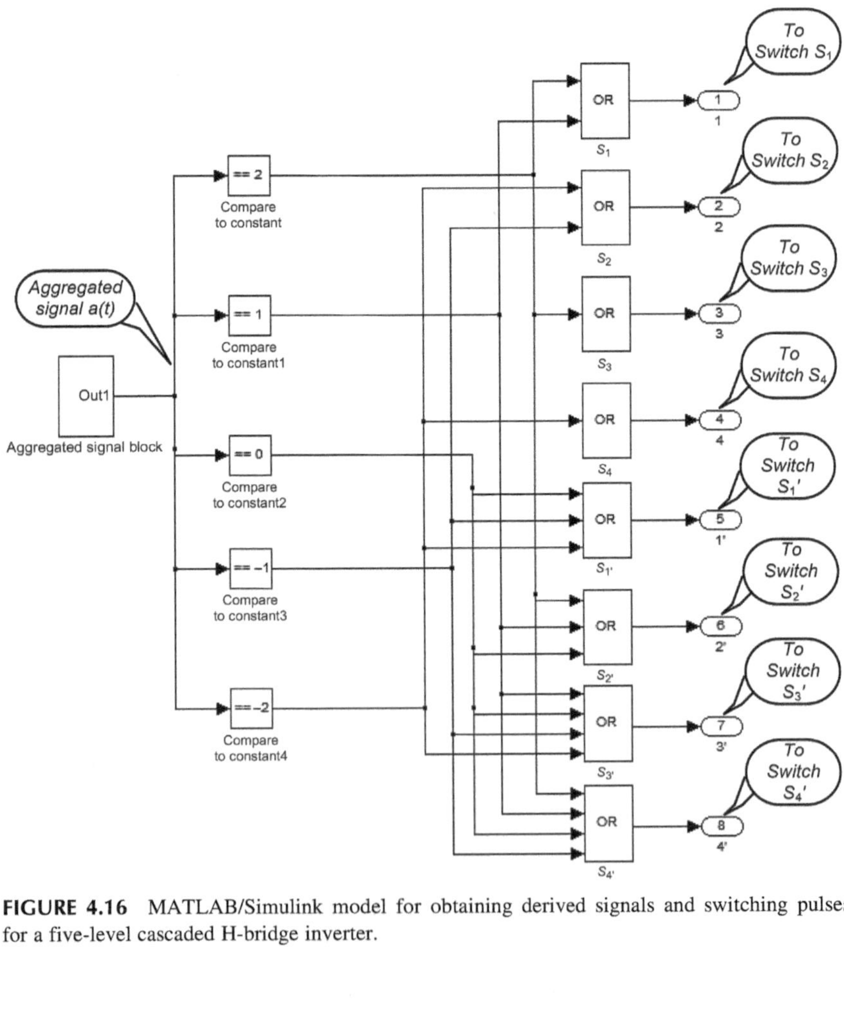

FIGURE 4.16 MATLAB/Simulink model for obtaining derived signals and switching pulses for a five-level cascaded H-bridge inverter.

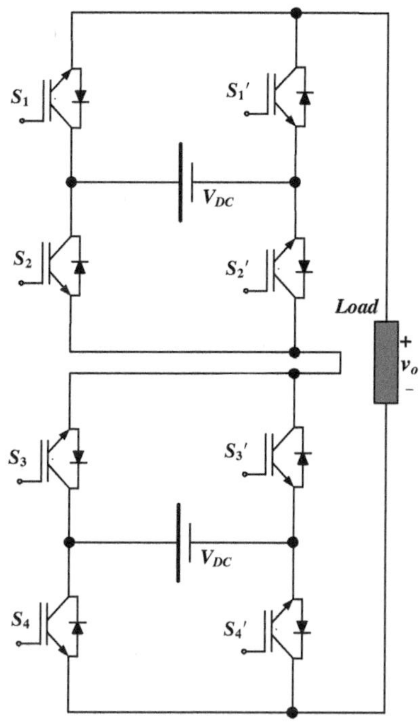

FIGURE 4.17 Power circuit of a five-level cascaded H-bridge topology.

TABLE 4.1 Look-up Table for Five-Level Cascaded H-Bridge Inverter

Level of Aggregated Signal $a(t)$	Output Voltage Level $v_o(t)$	Switches in ON State (Other Switches Remain OFF)
2	$2V_{DC}$	S_1, S_2', S_3, S_4'
1	V_{DC}	S_1, S_2', S_3', S_4'
0	0	S_1', S_2', S_3', S_4'
−1	$-V_{DC}$	S_1', S_2, S_3', S_4'
−2	$-2V_{DC}$	S_1', S_2, S_3', S_4

waveform, namely $\pm 2V_{DC}$, $\pm V_{DC}$, and 0. Each level in the aggregated signal *a(t)* corresponds to an "expected" level in the output voltage waveform (which, in turn, requires a specific switching combination). These one-to-one relationships are summarised in a look-up table. Table 4.1 is a look-up table for the case of a five-level cascaded H-bridge inverter and signal *a(t)*. It can be appreciated here that configuring the look-up table provides the flexibility to utilize the derived signals for any topological structure.

As the look-up table is formulated, the switching pulses for a given switch can be obtained by "OR"ing the derived signals corresponding to the level(s) for which the switch needs to conduct, as shown in Fig. 4.16.

4.4.3 Simulation of a Five-Level Cascaded H-Bridge Inverter

A MATLAB/Simulink model for simulation of a five-level cascaded H-bridge inverter is shown in Fig. 4.18, consisting of the power circuit, control circuit, and load along with measurement blocks. Input voltage sources are taken as $V_{DC} = 24$ V and an RL load with $R = 7.62\Omega$ and $L = 3.25$ mH is placed at the inverter terminals. Switching pulses for various switches are shown in Fig. 4.19, while the output voltage waveform and load current waveform along with their respective harmonic profiles are shown in Fig. 4.20. The output voltage is a five-level PWM waveform in equal steps of 24 V, while the current waveform indicates the inductive nature of the load. Thus, satisfactory performance of the model can be observed.

4.4.4 Experimental Implementation of a Five-Level Cascaded H-Bridge Inverter

A laboratory prototype of a five-level inverter based on cascaded H-bridge topology has been developed, to verify UCS. For the power circuit part, a MOSFET power module with switches IRF460 with suitable gate driver circuits is employed. To obtain control pulses, a real-time controller dSPACE DS1103 is interfaced with the MATLAB/Simulink model through Real Time Workshop. A schematic diagram of the layout is illustrated in Fig. 4.21.

The switching pulses obtained from the digital controller are shown in Fig. 4.22. Also, the load voltage and load current are shown along with their respective harmonic profiles in Fig. 4.23. A close similarity between the simulation and hardware results can be observed, thereby verifying the efficacy of UCS.

4.5 IMPLEMENTATION OF UCS IN SOME RECENTLY PROPOSED TOPOLOGIES

In Section 4.4, complete implementation of a cascaded H-bridge five-level inverter with the help of a proposed control scheme was demonstrated through simulations and experimentation. In this section three simulation studies are presented for three recent topologies, using UCS to demonstrate its universal appeal.

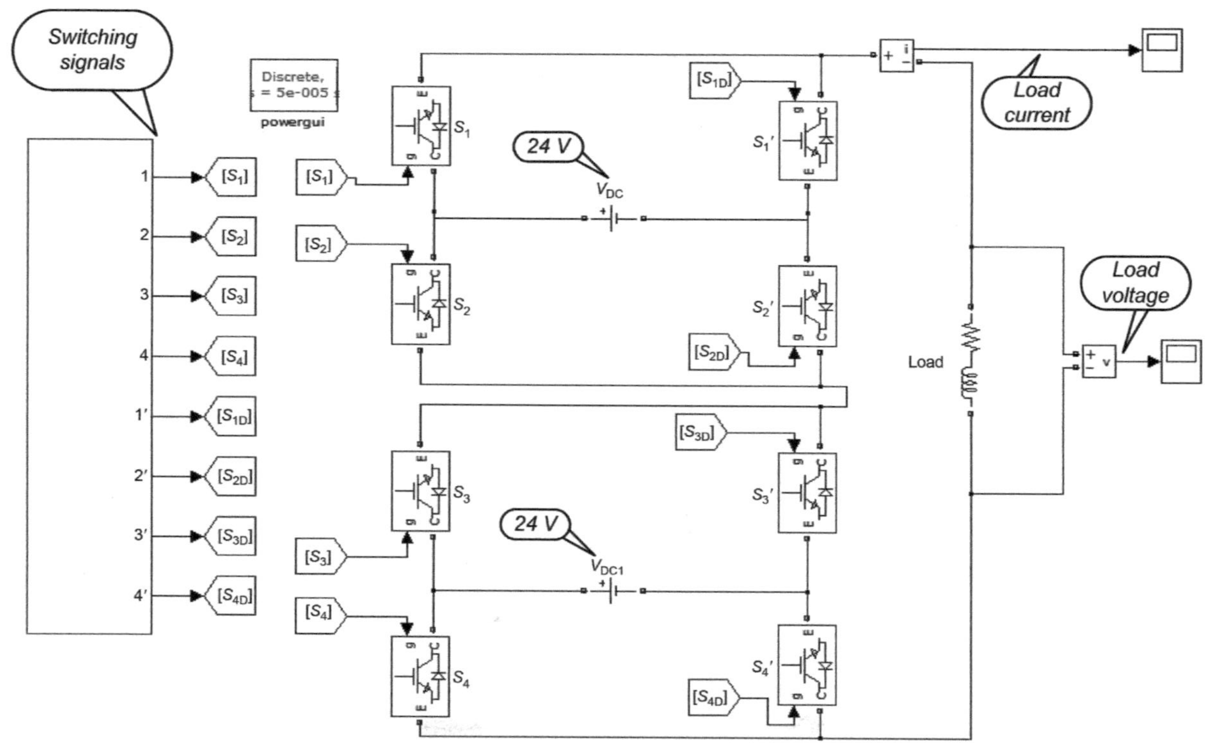

FIGURE 4.18 MATLAB/Simulink model for a five-level cascaded H-bridge inverter controlled by the proposed scheme.

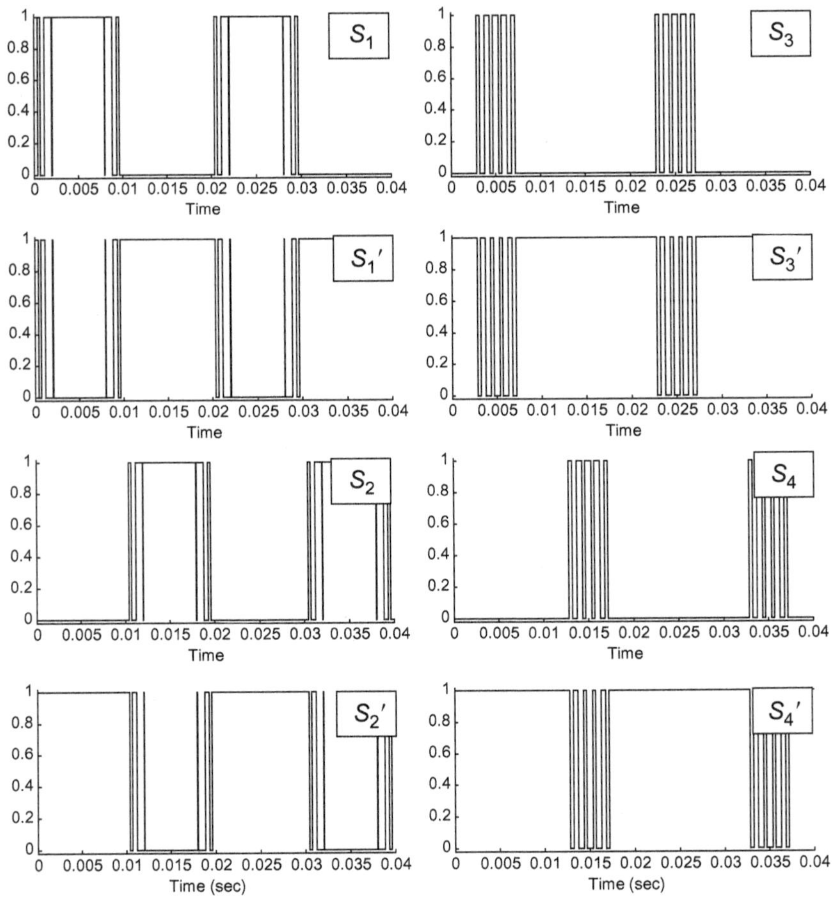

FIGURE 4.19 Switching pulses for the switches in a five-level cascaded H-bridge inverter.

4.5.1 Implementation for 2SELG-Based MLI

As discussed in Section 3.3.9 in Chapter 3, Advent of New Topologies, a cascaded MLI using series connection of sub-MLIs was proposed in Reference [24], with specific consideration of the reduction in the number of switching devices and high-voltage applications. The topology consists of nonisolated DC input sources and a mix of bidirectional-conducting-unidirectional-blocking and bidirectional-conducting-bidirectional-blocking switches. The power circuit of a single-phase nine-level inverter based on the topology proposed in Reference [24] is shown in Fig. 4.24, the topology being referred to as 2SELG-MLI henceforth.

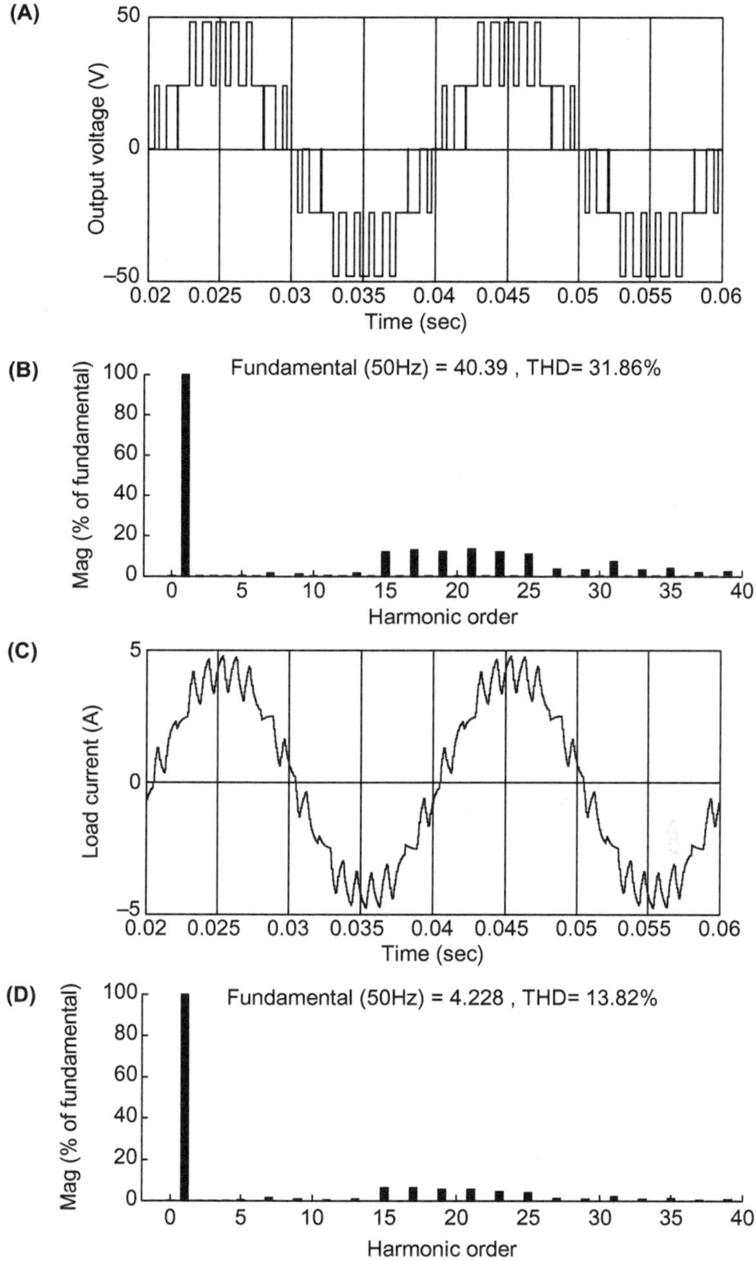

FIGURE 4.20 (A) Five-level output voltage (simulated) and (B) its harmonic profile. (C) Load current waveform (simulated) and (D) its harmonic profile.

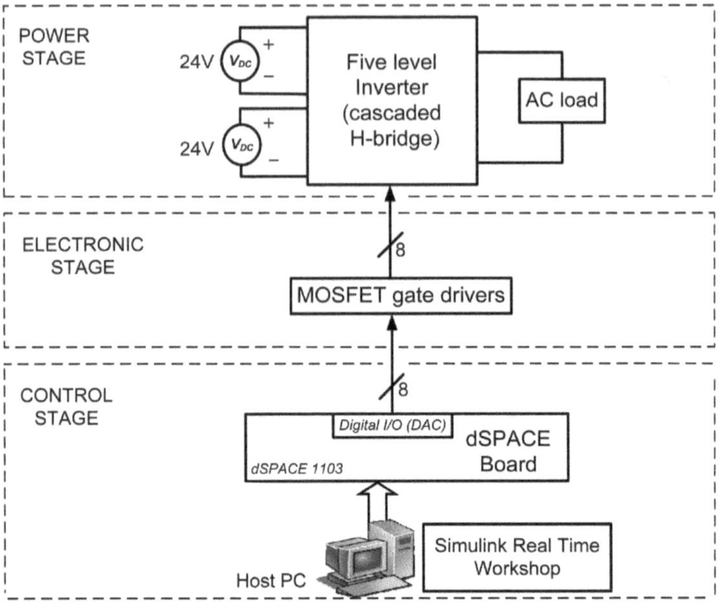

FIGURE 4.21 Schematic diagram of the laboratory prototype of a five-level cascaded H-bridge inverter.

In this subsection, the proposed control scheme is employed for modulating 2SELG-MLI using a SHE technique as described in Section 4.2.3. A nine-level aggregated signal that would be obtained is shown in Fig. 4.25. The signal has nine levels from −4 to +4 in steps of 1. For simulation study, the input sources are taken as $V_1 = V_2 = V_3 = V_4 = V_{DC} = 100$ V. Table 4.2 is an appropriate look-up table.

Based on this look-up table and the modulation scheme, the topology is simulated in MATLAB/Simulink environment, a load with $R = 2\Omega$ and $L = 5$ mH is considered. Load voltage and load current waveforms are shown in Fig. 4.26A and 4.26B, respectively. The load voltage is a multistepped waveform with nine equal steps of 100 V each. The harmonic spectrum of the load voltage is shown in Fig. 4.26C and it can be observed that targeted lower-order harmonics (namely third, fifth, and seventh) are effectively eliminated from the spectrum.

4.5.2 Implementation for Switched Series/Parallel Sources-Based MLI

Hinago et al. [21] have proposed a topology which uses a single source and switched series/parallel capacitors to obtain a multilevel waveform. A single-phase seven-level structure of the topology is shown in

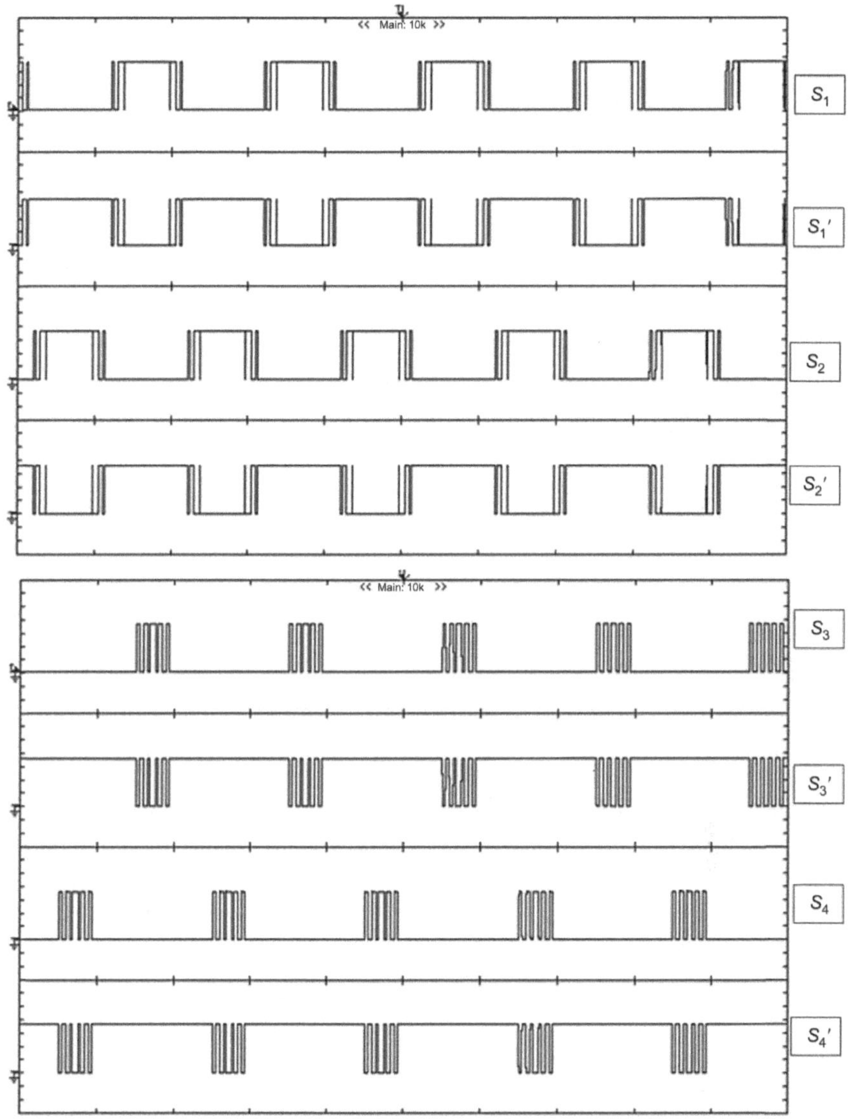

FIGURE 4.22 Experimentally obtained switching pulses for a five-level inverter.

Fig. 4.27, henceforth to be referred to as Switched Series/Parallel Sources-Based MLI (SSPS-MLI).

An aggregated signal for seven-level output using triangular waveforms of 40-kHz frequency and sinusoidal reference of 1-kHz frequency (values as used in Reference [21]) with APOD configuration as shown in Fig. 4.28.

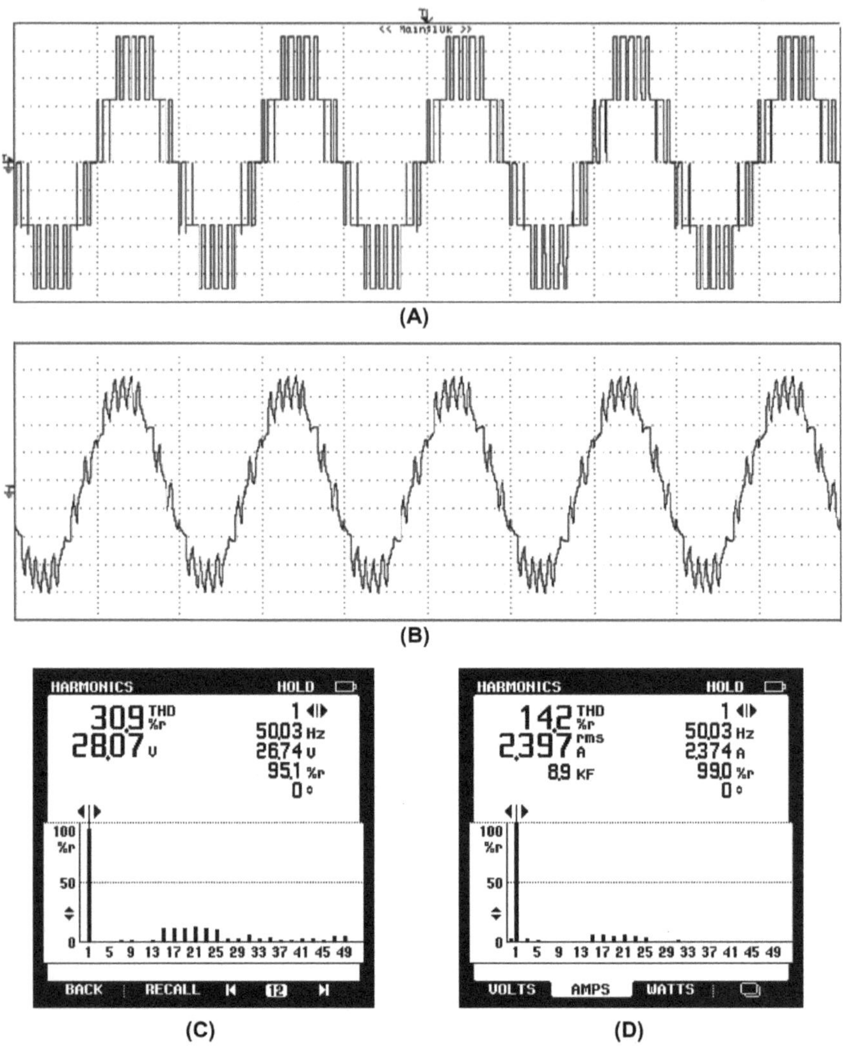

FIGURE 4.23 Experimental results for a five-level inverter based on cascaded H-bridge topology: (A) load voltage; (B) load current; (C) harmonic profile of load voltage; and (D) harmonic profile of load current.

The topology provides switching combinations for parallel connection of capacitors with the DC source for maintaining their state-of-charge and series connection to obtain a boosted output. The proposed control scheme can be employed in such a way that for output voltage zero and V_{DC}, the capacitors are kept in parallel with the source V_{DC}. Table 4.3 is a look-up table with this concept.

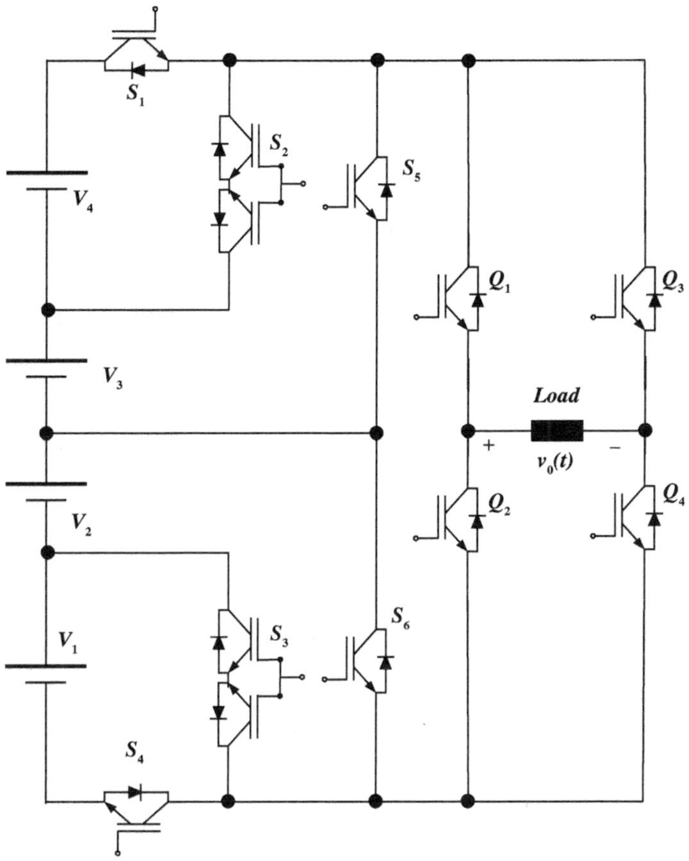

FIGURE 4.24 Power circuit of a nine-level inverter based on the topology proposed in Reference [24].

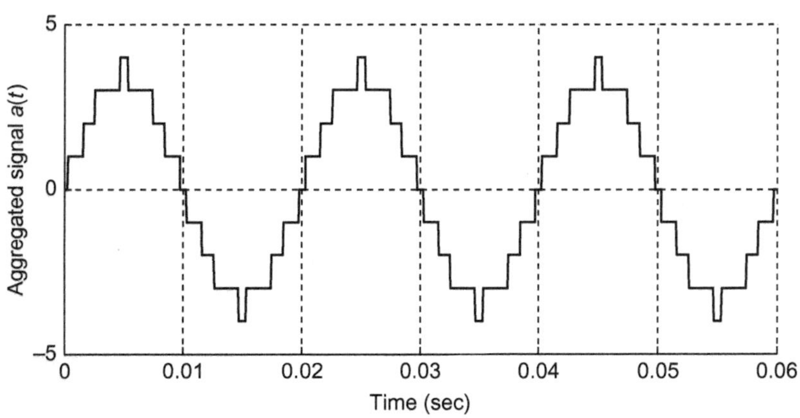

FIGURE 4.25 Aggregated signal *a(t)* for SHE-based modulation used for 2SELG-MLI.

TABLE 4.2 Look-up Table for a Nine-Level Inverter Based on 2SELG-MLI

Level of Aggregated Signal *a(t)*	Output Voltage Level $v_o(t)$	Switches in ON State (Other Switches Remain OFF)
4	$4V_{DC}$	S_1, S_4, Q_1, Q_4
3	$3V_{DC}$	S_2, S_4, Q_1, Q_4
2	$2V_{DC}$	S_4, S_5, Q_1, Q_4
1	V_{DC}	S_2, S_6, Q_1, Q_4
0	0	S_5, S_6, Q_1, Q_4
−1	$-V_{DC}$	S_2, S_6, Q_2, Q_3
−2	$-2V_{DC}$	S_4, S_5, Q_2, Q_3
−3	$-3V_{DC}$	S_2, S_4, Q_2, Q_3
−4	$-4V_{DC}$	S_1, S_4, Q_2, Q_3

Simulations are carried out with $V_{DC} = 100$ V, $C_1 = C_2 = 712$ μF, and an inductive load with $R = 400\ \Omega$ and $L = 4.89$ mH. Load voltage and load current waveforms are shown in Figs. 4.29 and 4.30, respectively. For the single-source DC input of 100 V, the load voltage is a boosted waveform with a peak of 300 V and seven steps of 100 V each. It can be observed that the desirable output voltage is obtained while the capacitors get properly charged to attain the required voltage levels at the output, as can be observed from the capacitor voltage waveforms shown in Figs. 4.31A and 4.31B.

4.5.3 Implementation for Reversing Voltage Topology

So far, the proposed scheme has been employed only for single-phase MLI structures. In this section, the scheme is shown for a three-phase five-level inverter based on the topology proposed in Reference [35]. Najafi and Yatim [35] proposed an interesting topology which significantly reduces the device count, especially for a three-phase system as compared to other popular topologies. A five-level three-phase structure is shown in Fig. 4.32, which is henceforth referred to as reversing voltage (RV) topology. This structure is simulated with a third harmonic injection method as described in Section 4.2.2 (Point 2). For each leg a five-level aggregated signal would be obtained and, accordingly, a look-up table for leg "*a*" is shown as in Table 4.4.

For a simulation study, each of the two input DC voltages is taken as V_{DC} and a three-phase inductive load with $R = 2\Omega$ and $L = 5$ mH in each phase is considered. The phase voltages as obtained at the output terminals are shown in Fig. 4.33. Each phase voltage has five levels in equal steps of 100 V and they

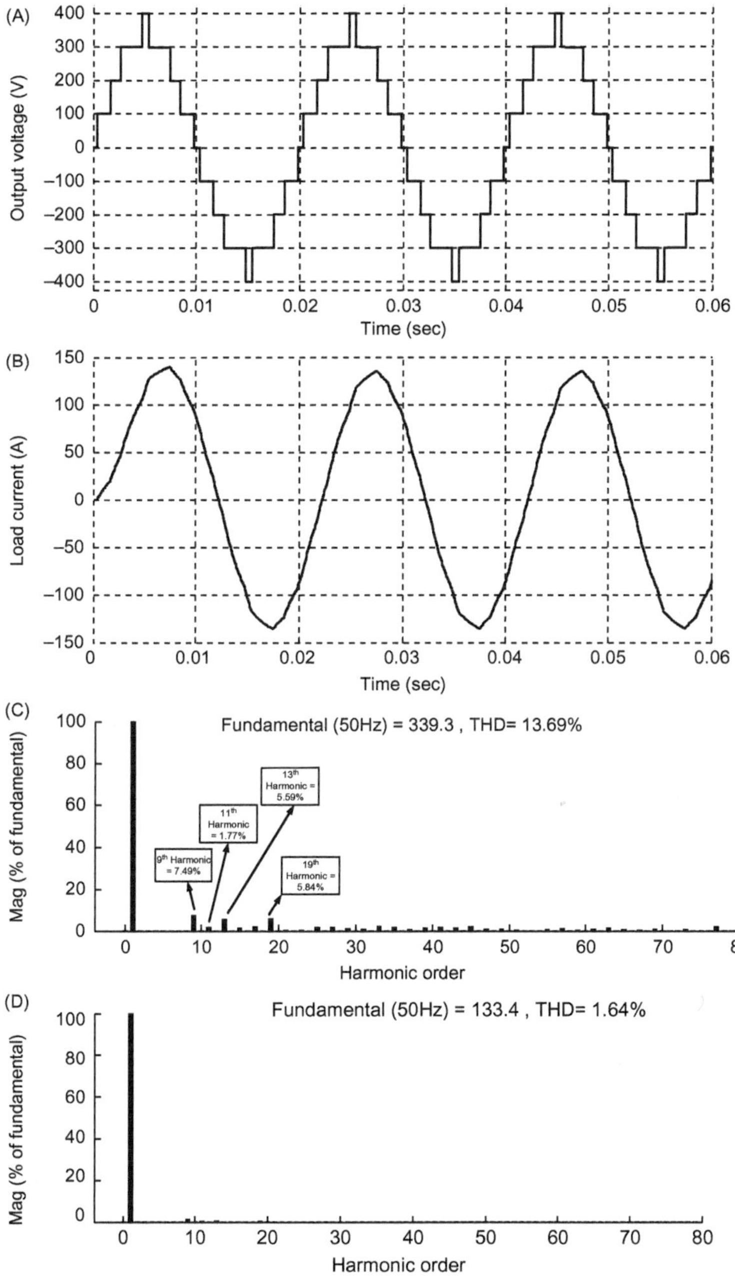

FIGURE 4.26 Simulated waveforms for 2SELG-MLI: (A) load voltage; (B) load current; (C) harmonic spectrum of load voltage; and (D) harmonic spectrum of load current.

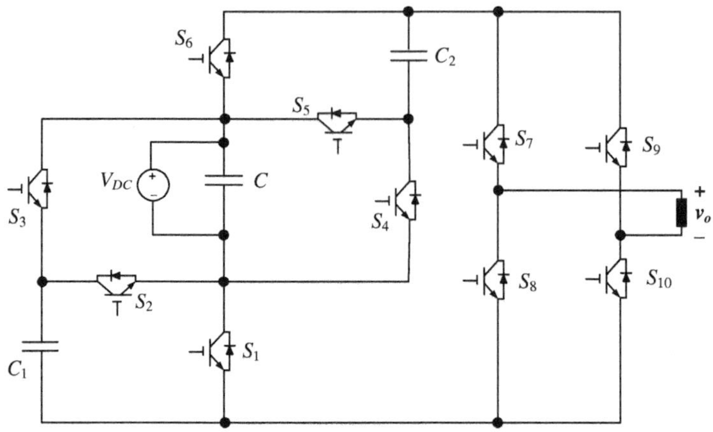

FIGURE 4.27 Power circuit of a seven-level inverter based on the topology proposed in Reference [21].

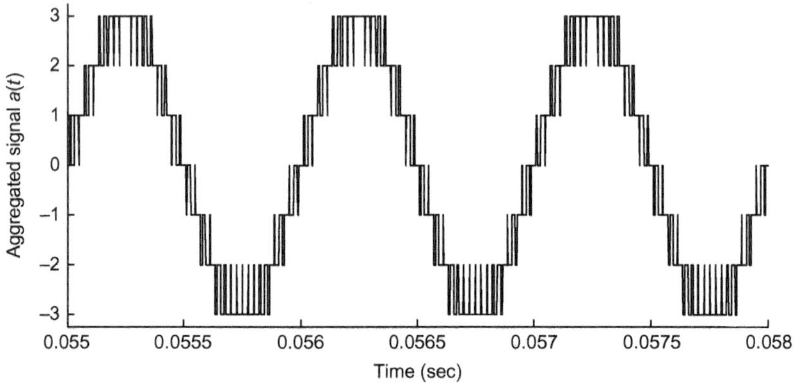

FIGURE 4.28 Aggregated signal *a(t)* for APOD-based modulation used for SSPS-MLI.

TABLE 4.3 Look-up Table for Seven-Level Inverter Based on SSPS-MLI

Level of Aggregated Signal *a(t)*	Corresponding Voltage Output Level	Switches in ON State
3	$+3V_{DC}$	S_2, S_5, S_7, S_{10}
2	$+2V_{DC}$	S_2, S_6, S_7, S_{10}
1	$+V_{DC}$	S_1, S_3, S_4, S_6, S_7, S_{10}
0	0	S_1, S_3, S_4, S_6, S_7, S_9
−1	$-V_{DC}$	S_1, S_3, S_4, S_6, S_8, S_9
−2	$-2V_{DC}$	S_2, S_6, S_8, S_9
−3	$-3V_{DC}$	S_2, S_5, S_8, S_9

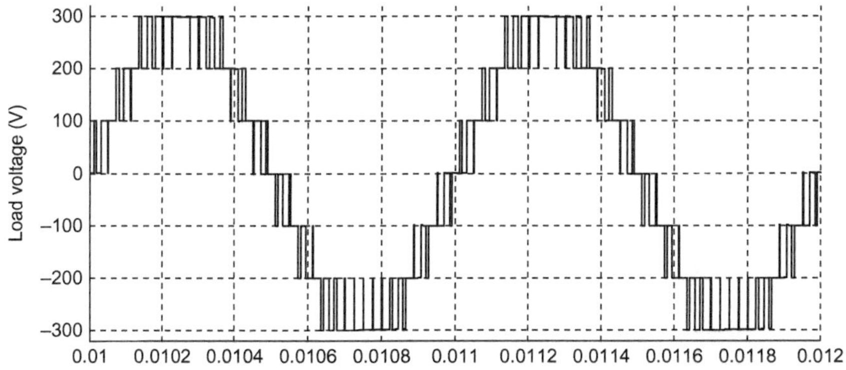

FIGURE 4.29 Output voltage waveform for SSPS-MLI.

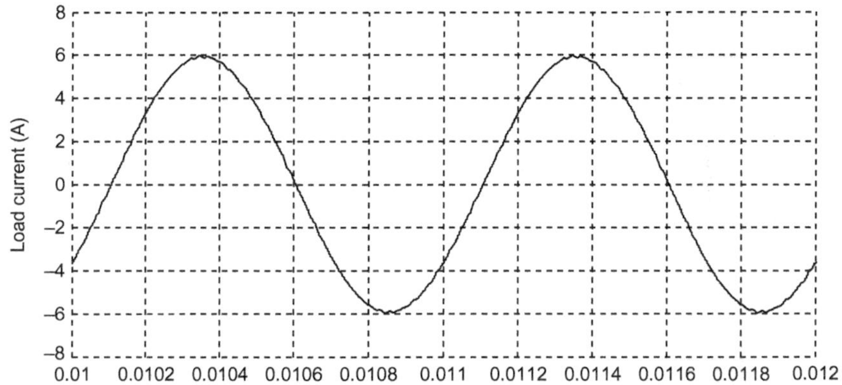

FIGURE 4.30 Load current waveform for SSPS-MLI.

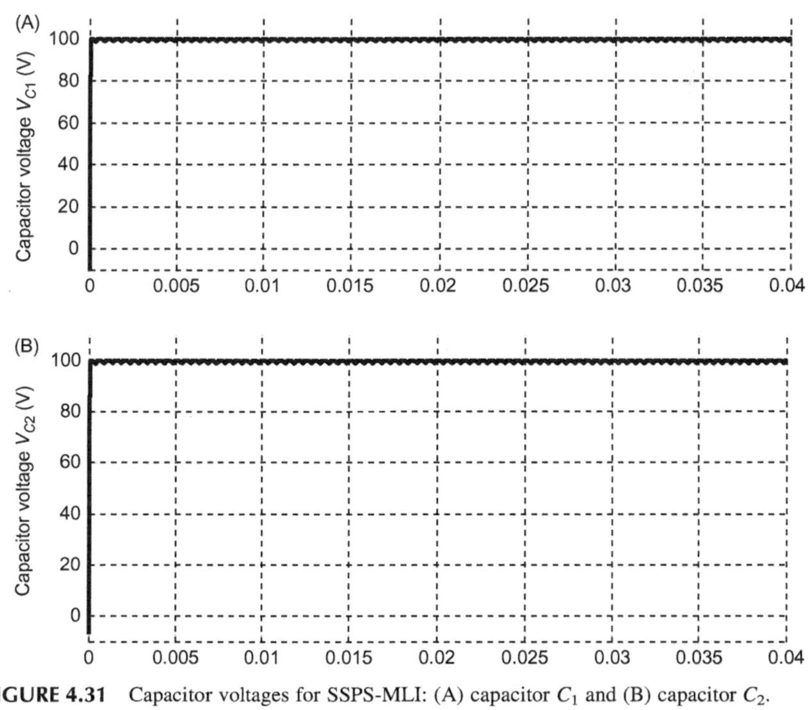

FIGURE 4.31 Capacitor voltages for SSPS-MLI: (A) capacitor C_1 and (B) capacitor C_2.

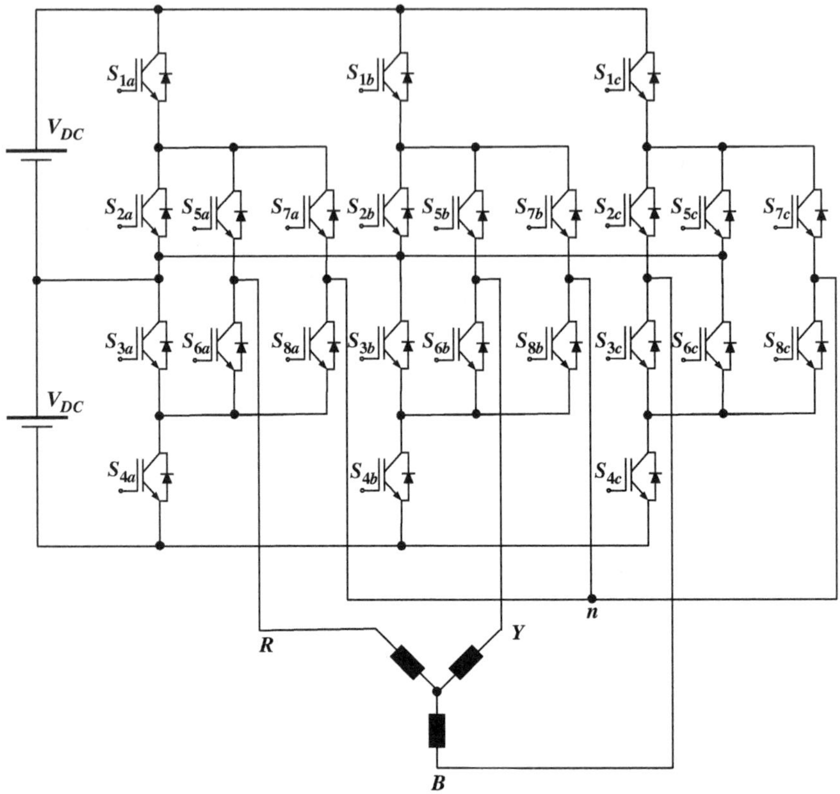

FIGURE 4.32 Power circuit of a three-phase five-level inverter based on the topology proposed in Reference [35].

TABLE 4.4 Look-up Table for Leg "a" of a Five-Level Inverter Based on RV Topology

Level of Aggregated Signal $a_r(t)$	Output Voltage Level $v_{Rn}(t)$	Switches in ON State (Other Switches Remain OFF)
2	$2V_{DC}$	S_{1a}, S_{4a}, S_{5a}, S_{8a}
1	V_{DC}	S_{2a}, S_{4a}, S_{5a}, S_{8a}
0	0	S_{2a}, S_{3a}, S_{5a}, S_{7a}
−1	$-V_{DC}$	S_{2a}, S_{4a}, S_{6a}, S_{7a}
−2	$-2V_{DC}$	S_{1a}, S_{4a}, S_{6a}, S_{7a}

are phase-shifted from one another by an angle of 120 degrees. Similarly, the line voltages are shown in Fig. 4.34, along with the harmonic spectrum of one line-voltage waveform. It can be seen that the line voltages comprise nine levels in equal steps of 50 V, as expected. Also, they are phase-shifted amongst them by 120 degrees. Load current waveforms are shown in Fig. 4.35.

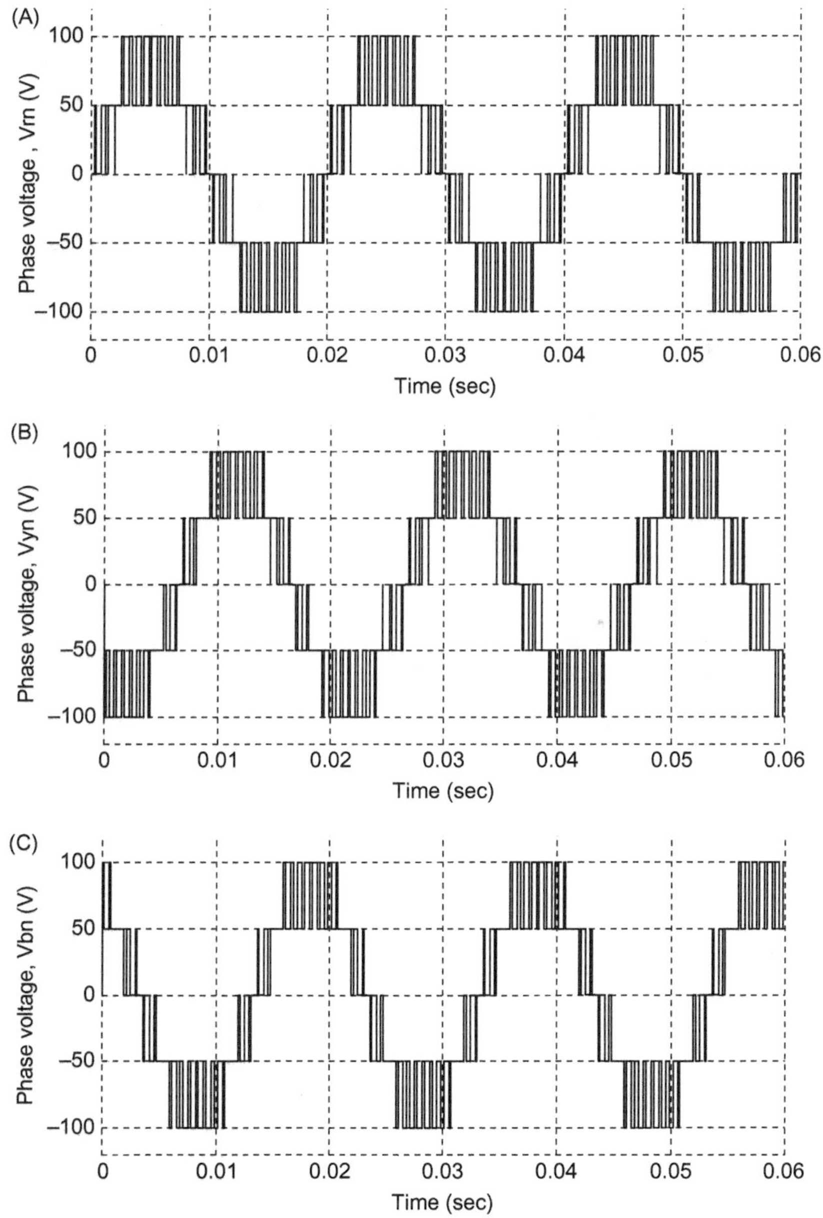

FIGURE 4.33 Three-phase line to neutral voltage waveforms for RV topology.

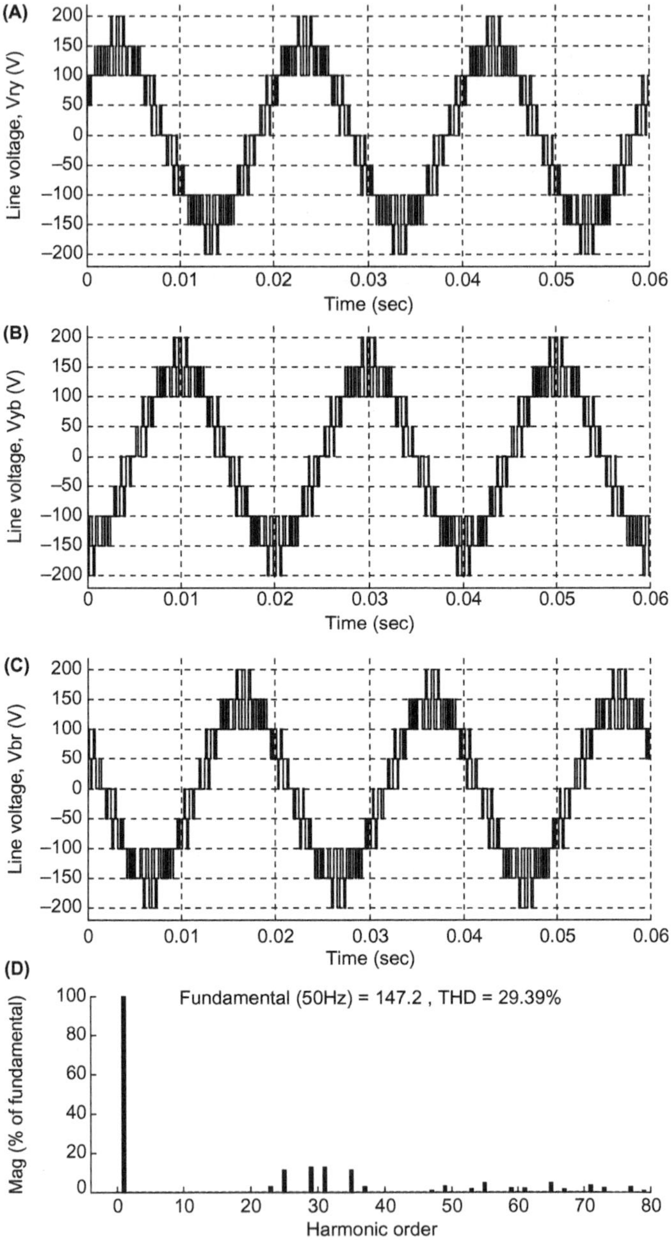

FIGURE 4.34 (A–C) Simulated line voltage waveforms for RV topology and (D) harmonic spectrum for $v_{ry}(t)$.

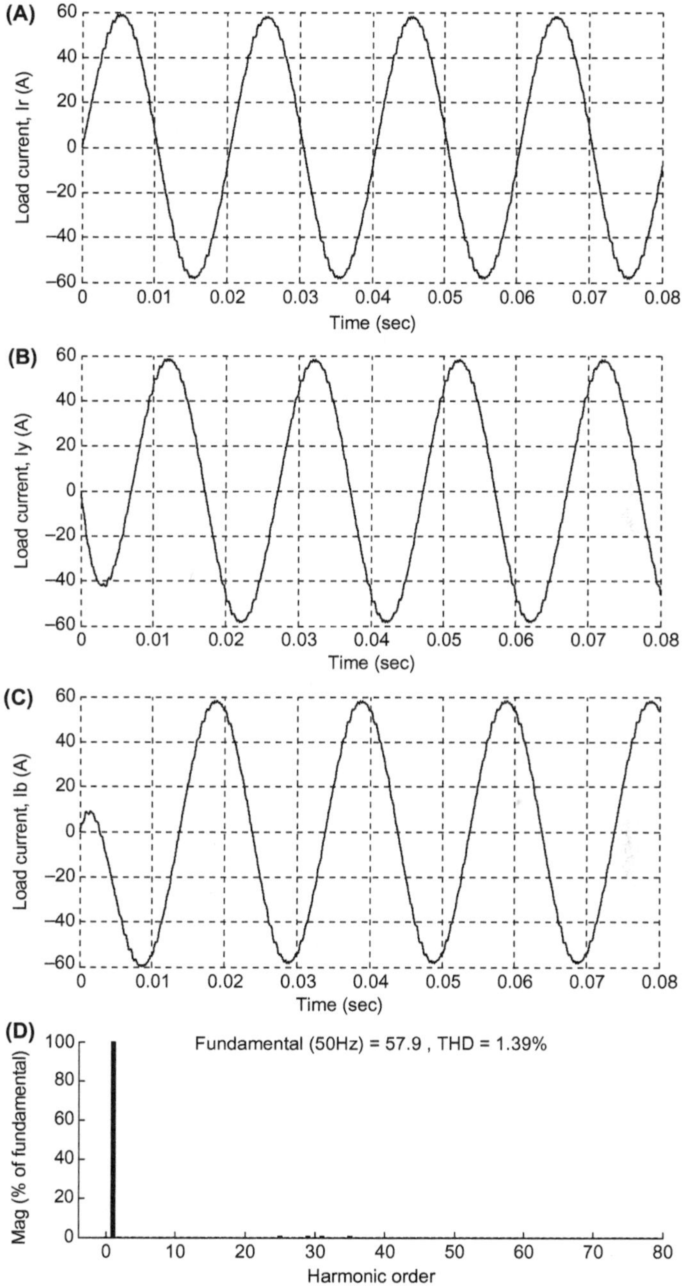

FIGURE 4.35 (A–C) Simulated line current waveforms for RV topology and (D) harmonic spectrum for $i_r(t)$.

4.6 SUMMARY

While space vector-based methods become cumbersome for modulating MLIs with an increase in the number of output levels, voltage-level-based modulation schemes are popular for such purposes. As newer topologies continue to emerge, a unified scheme is developed and discussed in this chapter so that any given topology can be modulated with any given voltage-level-based method to evaluate the performance of that topology. A mathematical description of the scheme is described. To verify the efficacy of the proposed scheme, three different genres of topologies are simulated with three distinct methods using the scheme. It is observed that the UCS can be employed for any topology with satisfactory performance.

REFERENCES

[1] D.G. Holmes, T. Lipo, IEEE Press Pulse Width Modulation for Power Converters: Principles and Practice, John Wiley, Hoboken, NJ, 2003.

[2] G. Carrara, S. Gardella, M. Marchesoni, R. Salutari, G. Sciutto, A new multilevel PWM method: a theoretical analysis, IEEE Trans. Pow. Electron. 7 (3) (1992) 497–507.

[3] B.P. McGrath, D.G. Holmes, Multicarrier PWM strategies for multilevel inverters, IEEE Trans. Industrial Electron. 49 (4) (2002) 858–867.

[4] A.M. Massoud, S.J. Finney, B.W. Williams, Control techniques for multilevel voltage source inverters, Power Electronics Specialist Conference, 2003. PESC '03. 2003 IEEE 34th Annual, vol. 1, pp. 171–176, June 15–19, 2003.

[5] C. Rech, J.R. Pinheiro, Impact of hybrid multilevel modulation strategies on input and output harmonic performances, IEEE Trans. Pow. Electron. 22 (3) (May 2007) 967–977.

[6] L.M. Tolbert, F.Z. Peng, T.G. Habetler, Multilevel PWM methods at low modulation indices, IEEE Trans. Pow. Electron. 15 (4) (2000) 719–725.

[7] Z. Haoran, A. Jouanne, D. Shaoan, A.K. Wallace, F. Wang, Multilevel inverter modulation schemes to eliminate common-mode voltages, IEEE Trans. Industry Appl. 36 (6) (2000) 1645–1653.

[8] N. Celanovic, D. Boroyevich, A fast space-vector modulation algorithm for multilevel three-phase converters, IEEE Trans. Industry Appl. 37 (2) (2001) 637–641.

[9] A.K. Gupta, A.M. Khambadkone, A space vector PWM scheme for multilevel inverters based on two-level space vector PWM, IEEE Trans. Industrial Electron. 53 (5) (2006) 1631–1639.

[10] A.K. Gupta, A.M. Khambadkone, A general space vector PWM algorithm for multilevel inverters, including operation in overmodulation range, IEEE Trans. Pow. Electron. 22 (2) (2007) 517–526.

[11] S. Wei, B. Wu, F. Li, C. Liu, A general space vector PWM control algorithm for multilevel inverters, Applied Power Electronics Conference and Exposition, 2003. APEC '03. Eighteenth Annual IEEE, vol. 1, pp. 562–568, February 9–13, 2003.

[12] J.I. Leon, S. Vazquez, A.J. Watson, L.G. Franquelo, P.W. Wheeler, J.M. Carrasco, Feed-forward space vector modulation for single-phase multilevel cascaded converters with any DC voltage ratio, IEEE Trans. Industrial Electron. 56 (2) (2009) 315–325.

[13] J. Rodriguez, L. Moran, P. Correa, C. Silva, A vector control technique for medium-voltage multilevel inverters, IEEE Trans. Industrial Electron. 49 (4) (2002) 882–888.

[14] M. Saeedifard, P. Barbosa, P. Steimer, Operation and control of a hybrid seven-level converter, IEEE Trans. Pow. Electron. 27 (2) (2012) 652–660.

[15] N.A. Rahim, M.M. Elias, W. Hew, Transistor-clamped H-bridge based cascaded multilevel inverter with new method of capacitor voltage balancing, IEEE Trans. Industrial Electron. 60 (8) (2013) 2943–2956.

[16] P. Lezana, C. Silva, R. Aceiton, Phase disposition PWM implementation for an hybrid multicell converter, IEEE Trans. Industrial Electron. 60 (5) (2013) 1936–1942.

[17] R. Naderi, A. Rahmati, Phase-shifted carrier PWM technique for general cascaded inverters, IEEE Trans. Pow. Electron. 23 (3) (2008) 1257–1269.

[18] F.Z. Peng, A generalized multilevel inverter topology with self voltage balancing, IEEE Trans. Industry Appl. 37 (2) (2001) 611–618.

[19] S. Gui-Jia, Multilevel DC link inverter. Industry Applications Conference, 2004. 39th IAS Annual Meeting. Conference Record of the 2004 IEEE, vol. 2, pp. 806–812, October 3–7, 2004.

[20] G. Ceglia, V. Guzman, C. Sanchez, F. Ibanez, J. Walter, M.I. Gimenez, A new simplified multilevel inverter topology for DC-AC conversion, IEEE Trans. Pow. Electron. 21 (5) (2006) 1311–1319.

[21] Y. Hinago, H Koizumi, A single-phase multilevel inverter using switched series/parallel DC voltage sources, IEEE Trans. Industrial Electron. 57 (8) (2010) 2643–2650.

[22] E. Babaei, A cascade multilevel converter topology with reduced number of switches, IEEE Trans. Pow. Electron. 23 (6) (2008) 2657–2664.

[23] Y. Ounejjar, K. Al-Haddad, A novel high energetic efficiency multilevel topology with reduced impact on supply network. Industrial Electronics, 2008. IECON 2008. 34th Annual Conference of IEEE, pp. 489–494, November 10–13, 2008.

[24] M.F. Kangarlu, E Babaei, A generalized cascaded multilevel inverter using series connection of submultilevel inverters, IEEE Trans. Pow. Electron. 28 (2) (2013) 625–636.

[25] Hochgraf, C.G., High voltage modular inverter. U.S. Patent 6072707, 2000.

[26] Y.S. Lai, F.S Shyu, Topology for hybrid multilevel inverter, Electric Pow. Appl. IEE Proc. 149 (6) (2002) 449–458.

[27] K. Wang, L. Xu, Z. Zheng, Y. Li, Voltage balancing control of a four-level hybrid-clamped converter based on zero-sequence voltage injection using phase-shifted PWM, IEEE Trans. Pow. Electron. 31 (8) (2016) 5389–5399.

[28] S.J. Park, F.S. Kang, M.H. Lee, C.U. Kim, A new single-phase five-level PWM inverter employing a deadbeat control scheme, IEEE Trans. Pow. Electron. 18 (3) (2003) 831–843.

[29] J. Rodriguez, J.S. Lai, F.Z. Peng, Multilevel inverters: a survey of topologies, controls, and applications, IEEE Trans. Industrial Electron. 49 (2002) 724–738.

[30] L.G. Franquelo, J. Rodriguez, J.I. Leon, S. Kouro, R. Portillo, M.A.M. Prats, The age of multilevel converters arrives, Industrial Electron. Magaz. IEEE 2 (2) (2008) 28–39.

[31] J.K. Steinke, Switching frequency optimal PWM control of a three-level inverter, IEEE Trans. Pow. Electron. 7 (3) (1992) 487–496.

[32] E. Guan, P. Song, M. Ye, B. Wu, Selective harmonic elimination techniques for multilevel cascaded H-bridge inverters. Power Electronics and Drives Systems, 2005. PEDS 2005. International Conference on, vol. 2, pp. 1441–1446, 2005.

[33] S. Sirisukprasert, J.S. Lai, T.H. Liu, Optimum harmonic reduction with a wide range of modulation indexes for multilevel converters, IEEE Trans. Industrial Electron. 49 (4) (2002) 875–881.
[34] M. Perez, J. Rodriguez, J. Pontt, S. Kouro, Power distribution in hybrid multi-cell converter with nearest level modulation. Industrial Electronics, 2007. ISIE 2007. IEEE International Symposium on, pp. 736–741, June 4–7, 2007.
[35] E. Najafi, A. Yatim, Design and implementation of a new multilevel Inverter topology, IEEE Trans. Industrial Electron. 59 (11) (2012) 4148–4154.

Chapter 5

Multilevel Inverter Based on Bridge-Type Connected Sources

5.1 INTRODUCTION

The merit of a multilevel inverter is enhanced by increasing the number of output levels (as this results in better waveform quality and lower *dv/dt* stress on the load), but it inadvertently leads to an increased number of components. Researchers have presented many topologies to reduce the component count. These approaches can be broadly grouped as:

1. topological modifications [1–4];
2. employing asymmetric source configurations [5–7] or asymmetric ratios of capacitor voltages [8–10]; and
3. a combination of (1) and (2) [11].

It should also be mentioned that attempting to reduce the number of semiconductor switches in multilevel inverters involves one or more of the following compromises/challenges:

1. Increased power rating of semiconductor switches [12–14]
2. Increased number of sources [15]
3. Loss of modularity (as happens in asymmetric CHB topology) [14]
4. Reduced number of redundant states (as happens in asymmetric configurations) [11]
5. Complex modulation/control schemes [11]
6. Difficulty in obtaining charge balance control (especially in an open loop mode) [16].

In spite of these compromises, there remains a possibility of exploring topologies which can maximize the number of levels with given input DC levels. In this chapter, a new multilevel inverter topology is presented which is capable of synthesizing all possible additive and subtractive combinations of input DC levels. This chapter will help the reader to understand how a topology can be conceptualized and the manner in which it should be systematically presented and analyzed. Moreover, the state selection procedure as described previously is used to verify the topology with the help of simulation and experimental studies.

Multilevel Inverters. DOI: https://doi.org/10.1016/B978-0-12-812448-2.00005-2

5.2 CONCEPTUALIZATION OF TOPOLOGY

The concept adopted for the topology is that the inverter should be capable of synthesizing all possible additive and subtractive combinations of input DC levels at the output waveform.

Examples of such combinations include:

1. With a single input DC level

   If a single input DC level (represented by a DC source of voltage $V_{DC,1}$) is present, the possible voltage levels that can be synthesized are: $\pm V_{DC,1}$ and 0. Thus, the resultant inverter should be a three-level inverter. Such a structure is shown in Fig. 5.1, which is a standard single-phase full-bridge inverter, also known as an H-bridge inverter.
2. With two input DC levels

   If two input DC levels (represented by DC sources of voltages $V_{DC,1}$ and $V_{DC,2}$) are present, the possible combinations are:

   $$\pm V_{DC,1},\ \pm V_{DC,2},\ \pm(V_{DC,1} - V_{DC,2}),\ \pm(V_{DC,1} + V_{DC,2}) \text{ and } 0$$

   Thus, for two DC sources, there are a total of nine possible combinations and the resulting structure should produce nine levels, either distinct or redundant. Such a structure is shown in Fig. 5.2. Various switching combinations for these output levels are summarized in Table 5.1.

   It is also important to observe the interconnection between the DC sources through the power switches. A part of the structure, containing the two DC sources ($V_{DC,1}$ and $V_{DC,2}$) and interconnecting switches (S_2, S_2', S_3 and S_3') is shown in Fig. 5.3A and an equivalent structure is shown in Fig. 5.3B. As can be observed from Fig. 5.3B, the sources are interconnected in a "*bridge fashion*" and henceforth the topology is referred to as a "*bridge-connected sources–based multilevel inverter (BCS-MLI).*"
3. With three input DC levels

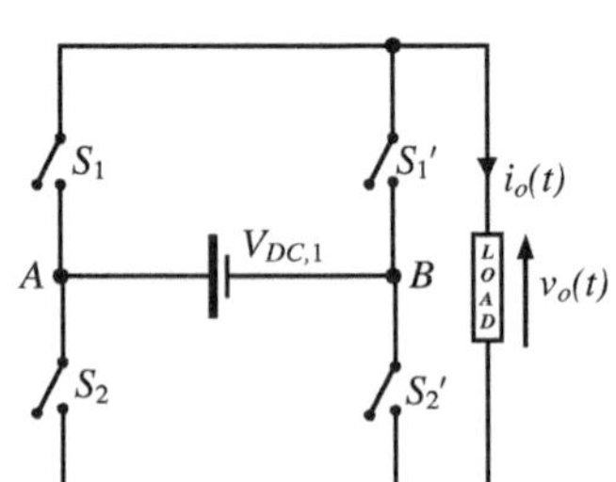

FIGURE 5.1 A full-bridge (or H-bridge) single-phase inverter.

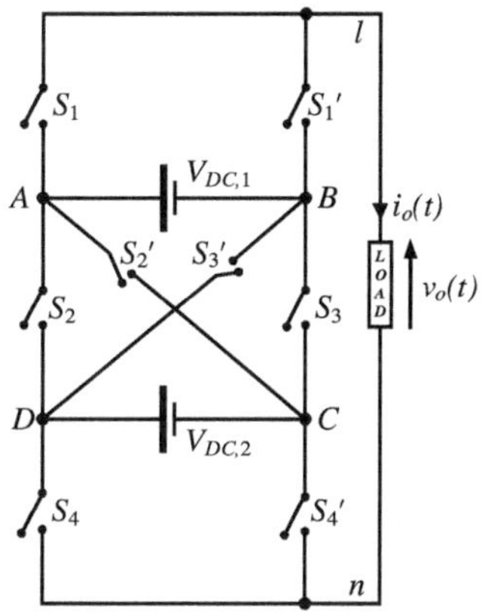

FIGURE 5.2 A topology for nine-level output.

TABLE 5.1 Switching States for Nine-Level Output

State	Output Voltage Level $v_0(t)$	Switch States (1 = *ON*, 0 = *OFF*)							
		S_1	S_2	S_3	S_4	S_1'	S_2'	S_3'	S_4'
1	$V_{DC,1}$	1	0	1	0	0	0	0	1
2	$V_{DC,1}$	1	0	0	1	0	0	1	0
3	$V_{DC,2}$	0	0	0	0	1	0	1	1
4	$V_{DC,2}$	1	1	0	0	0	0	0	1
5	$V_{DC,1} - V_{DC,2}$	1	0	1	1	0	0	0	0
6	$V_{DC,1} + V_{DC,2}$	1	0	0	0	0	0	1	1
7	0	0	0	1	0	1	0	0	1
8	0	0	0	0	1	1	0	1	0
9	0	1	1	0	1	0	0	0	0
10	0	1	0	0	0	0	1	0	1
11	$-V_{DC,1}$	0	1	0	1	1	0	0	0
12	$-V_{DC,1}$	0	0	0	0	1	1	0	1
13	$-V_{DC,2}$	0	0	1	1	1	0	0	0
14	$-V_{DC,2}$	1	0	0	1	0	1	0	0
15	$-V_{DC,1} + V_{DC,2}$	0	1	0	0	1	0	0	1
16	$-V_{DC,1} - V_{DC,2}$	0	0	0	1	1	1	0	0

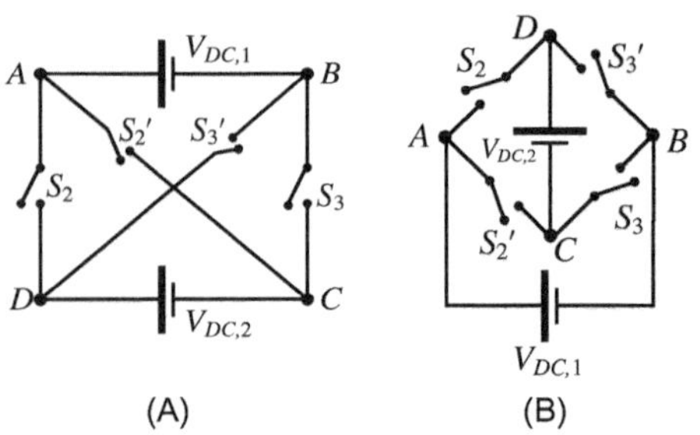

FIGURE 5.3 (A) Interconnection of input DC sources in the topology. (B) Equivalent interconnection of the input DC sources in the topology.

If three DC levels (represented by sources with voltages $V_{DC,1}$, $V_{DC,2}$, and $V_{DC,3}$) are available, the possible combinations are:

$$\begin{aligned}
&\pm V_{DC,1},\ \pm V_{DC,2},\ \pm V_{DC,3},\ \pm\left(V_{DC,1}-V_{DC,2}\right),\ \pm\left(V_{DC,1}+V_{DC,2}\right),\\
&\pm\left(V_{DC,1}-V_{DC,3}\right),\ \pm\left(V_{DC,1}+V_{DC,3}\right),\ \pm\left(V_{DC,2}-V_{DC,3}\right),\\
&\pm\left(V_{DC,2}+V_{DC,3}\right),\ \pm\left(V_{DC,1}+V_{DC,2}-V_{DC,3}\right),\\
&\pm\left(V_{DC,1}-V_{DC,2}+V_{DC,3}\right),\ \pm\left(V_{DC,1}-V_{DC,2}-V_{DC,3}\right),\\
&\pm\left(V_{DC,1}+V_{DC,2}+V_{DC,3}\right),\ \text{and } 0.
\end{aligned}$$

Thus, when three input DC levels are available, there are 27 additive and subtractive combinations possible, including a zero level.

4. With "*N*" number of input DC levels

A generalized structure can be obtained for "*N*" number of input DC levels. If "*N*" number of DC levels (represented by DC sources with voltages $V_{DC,p}$ [where $p = 1$ to N]) are present, then the number of possible combinations can be obtained by proceeding in the same manner as done for one, two, and three and taking "*m*" number of sources at a time, where "*m*" varies from 1 to "*N*."

The generalized expression for the number of levels can be obtained as:

$$\text{Maximum number of levels} = \left(\sum_{m=1}^{N} \frac{N!}{m\,!(N-m)\,!} 2^m\right) + 1 \tag{5.1}$$

The term "maximum" is used in the above expression because redundancy in the total number of states may decrease the actual number of levels in the output. The generalized structure of the BCS-MLI topology is shown in Fig. 5.4.

Eq. (5.1) indicates the possibility of synthesizing the maximum number of levels. However, the actual number of levels will depend on the source configuration. Various DC source configurations and the corresponding number of levels are discussed below:

1. "Unary" configuration consists of DC sources of equal voltages, i.e.,

$$V_{DC,1} = V_{DC,2} = V_{DC,3} = \ldots = V_{DC,N} = V_{DC} \tag{5.2}$$

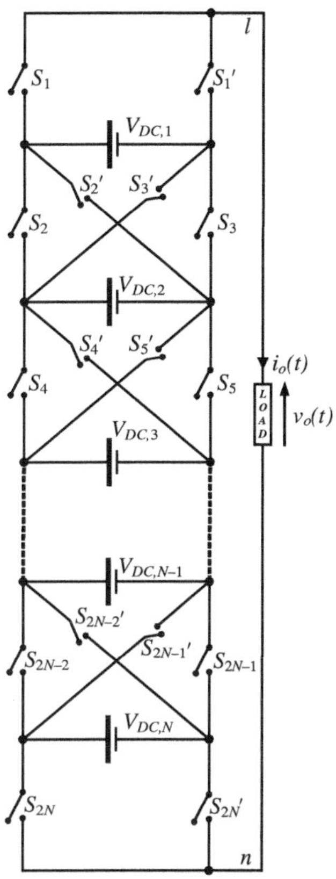

FIGURE 5.4 Generalized single-phase structure of BCS-MLI.

Unary configuration is also called symmetric source configuration. Such a configuration results in $(2N + 1)$ combinations of distinct levels.

2. "Binary" configuration consists of DC sources with a geometric progression with a factor of "2," i.e.,

$$\frac{V_{DC,1}}{V_{DC,2}} = \frac{V_{DC,2}}{V_{DC,3}} = \ldots = \frac{V_{DC,N-1}}{V_{DC,N}} = 2 \tag{5.3}$$

Such a configuration results in $(2^{(N+1)} - 1)$ combinations of distinct levels.

3. "Trinary" configuration consists of DC sources with a geometric progression with a factor of "3," i.e.,

$$\frac{V_{DC,1}}{V_{DC,2}} = \frac{V_{DC,2}}{V_{DC,3}} = \ldots = \frac{V_{DC,N-1}}{V_{DC,N}} = 3 \tag{5.4}$$

Such a configuration results in (3^N) combinations of distinct levels.

It is also noted here that in the BCS-MLI structure with "N" number of DC sources, the total number of switching states is determined as:

$$\text{Number of switching states} = 2^{2N} \tag{5.5}$$

5.2.1 Principle of Operation

In this section, the working principle of the BCS-MLI structure is explained with the help of a nine-level inverter for a single-phase operation.

A nine-level inverter with two DC sources $V_{DC,1}$ and $V_{DC,2}$ is shown in Fig. 5.2. The source configuration (unary, binary, or trinary) will dictate the actual number of levels in the output waveform obtained from the BCS-MLI. To obtain any desired voltage level, three power switches need to be ON simultaneously, as observed from Table 5.1. There are four pairs of switches. Switch pairs (S_1, S_1') and (S_4, S_4') operate in a complementary fashion. Switch pairs (S_2, S_2') and (S_3, S_3') are not complementary in the strict sense of the term but they operate in a way that the two members of a pair cannot be ON simultaneously, though they can remain OFF simultaneously. Thus, three switches (to remain ON) can be chosen from four possibilities in $^4C_3$ ($=4$) ways and each switch has two possibilities (either ON or OFF). Thus, the total number of states possible for the topology is $2^4 = 16$. All 16 states are shown in Fig. 5.5 along with the corresponding output voltage levels. It can be seen that in this single-phase configuration, the AC load is fed by nine levels $\pm V_{DC,1}$, $\pm V_{DC,2}$, $\pm$ ($V_{DC,1} - V_{DC,2}$), $\pm$ ($V_{DC,1} + V_{DC,2}$) and "0." In State 1, Switch S_4' conducts along with S_1 and S_3 so that the potential difference across the load is $V_{DC,1}$. Similarly, an output of ($V_{DC,1} - V_{DC,2}$) is obtained when S_1, S_3, and S_4 are ON. Various redundant states can also be noted. Voltage levels $\pm V_{DC,1}$ and $\pm V_{DC,2}$ have two redundant states each, which are (1, 2), (3, 4), (11, 12), and (13, 14), respectively. Zero level has four redundant states, namely, 7, 8, 9, and 10. It is evident that asymmetry in source configuration decreases the number of redundant states. For example, if $V_{DC,1} = V_{DC,2} = V_{DC}$, then with reference to Table 5.1, the output levels $\pm V_{DC}$ and 0 would have an increased number of redundant states. However, irrespective of the source configuration, the level $\pm$($V_{DC,1} + V_{DC,2}$) would have just one working state and no redundant state.

5.2.2 Output Voltage

Corresponding to a switch S_j, let T_j be a switching function, where $j = \{1, 2, 3, 4\}$, defined as:

$$T_j = 1, \text{ if } S_j \text{ is ON}$$

$$0, \text{ if } S_j \text{ is OFF} \tag{5.6}$$

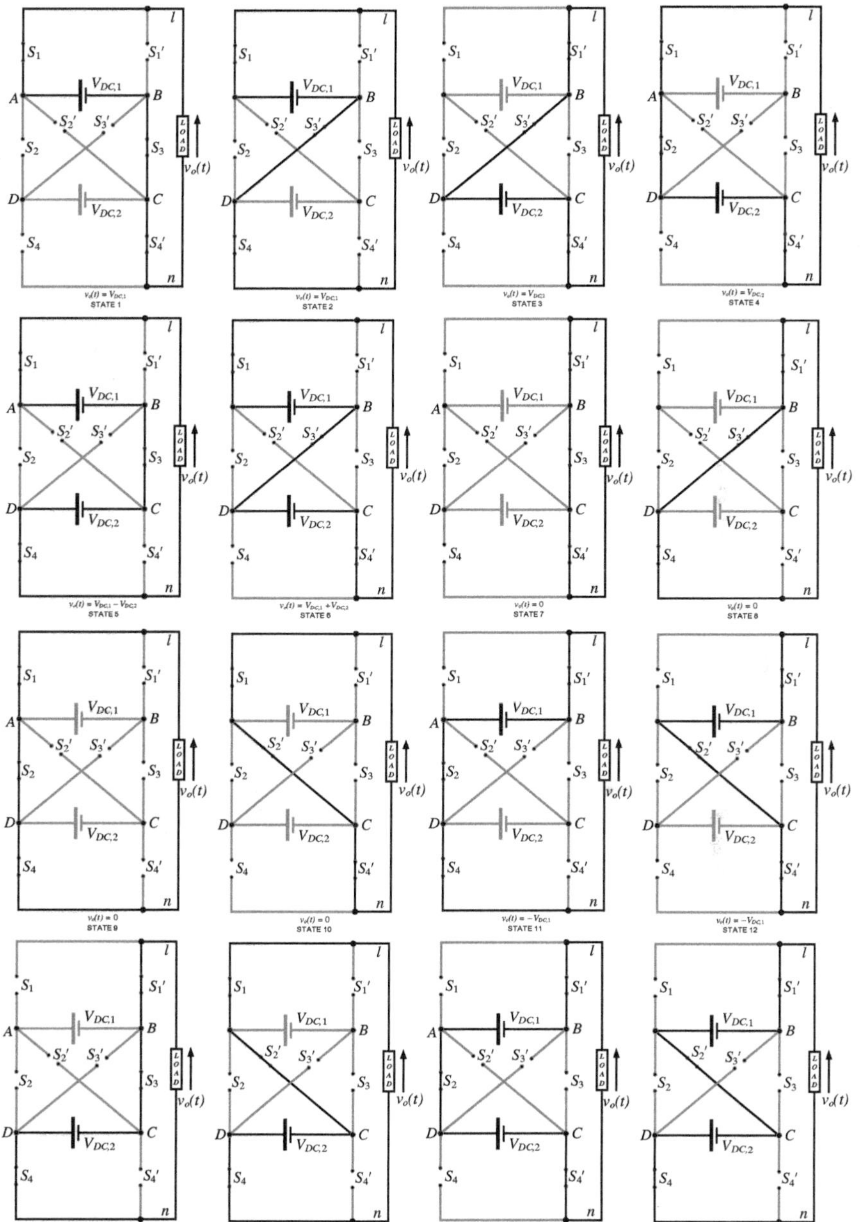

FIGURE 5.5 Working states for BCS-MLI with two input DC sources.

With reference to Fig. 5.2, output voltage $v_O(t)$ can be expressed in terms of switching functions as:

$$v_O(t) = -(1-T_1)V_{DC,1} + (1-T_2)\{-T_2'V_{DC,2} + T_3V_{DC,1} + T_3'(V_{DC,1} - V_{DC,2})\} + (1-T_4)V_{DC,2} \tag{5.7}$$

5.2.3 Voltage Across Blocking Switches

For BCS-MLI, semiconductor switches with different capabilities to block voltages are required. Assuming the case of an asymmetric topology with distinct sources $V_{DC,1}$ and $V_{DC,2}$ and $V_{DC,1} > V_{DC,2}$, the structure would require four switches (namely, S_1, S_1', S_2, and S_3) with minimum blocking voltage of $V_{DC,1}$, two switches (namely, S_4 and S_4') with minimum blocking voltage of $V_{DC,2}$, and two switches (namely, S_2' and S_3') with minimum blocking voltage of $(V_{DC,1} + V_{DC,2})$.

5.2.4 Power Switch Configuration

BCS-MLI-based DC-to-AC conversion can be realized with self-commutating power switches like MOSFETs and IGBTs. It is also important to note that the switches at positions S_2 and S_3 are necessarily required to be bidirectional-conducting-bidirectional-blocking (or "fully directional," sometimes simply called "bidirectional") switches otherwise their undesirable switching will take place. This is described with the help of Fig. 5.6 where

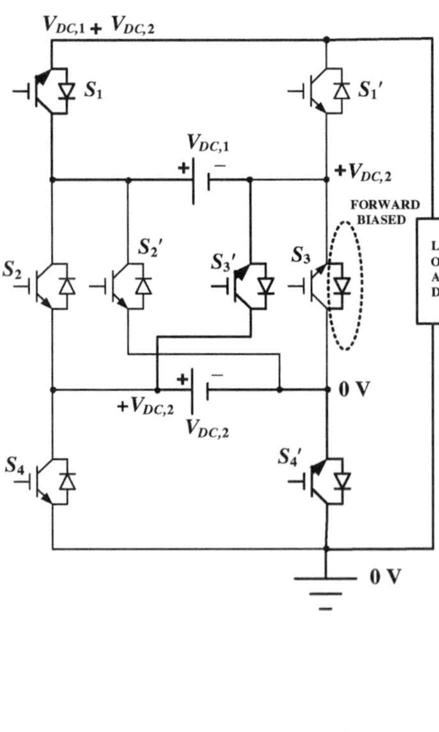

FIGURE 5.6 Pertaining to the absolute necessity of two "bidirectional" power switches.

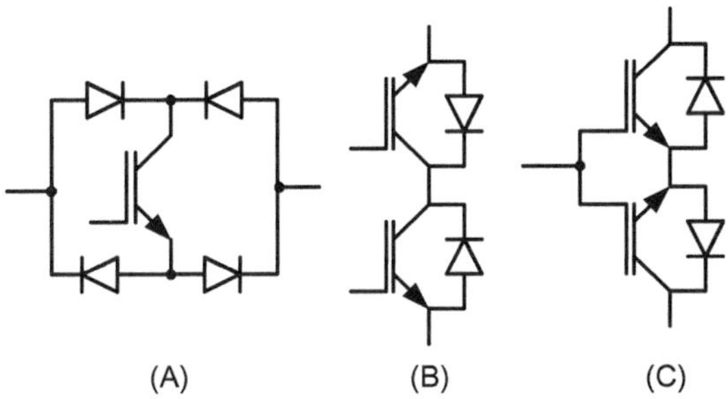

FIGURE 5.7 Various configurations for a bidirectional-conducting-bidirectional-blocking switch.

IGBTs (with antiparallel diodes) are shown at the S_2 and S_3 locations. When an output level ($V_{DC,1}$ + $V_{DC,2}$) is required and switches S_1, S_3', and S_4' are switched ON, then the antiparallel diode of Switch S_3 gets a forward-biasing potential difference of $V_{DC,2}$. Thus, it acts as an ON switch, thereby short-circuiting the source $V_{DC,2}$. A similar phenomenon happens when Switches S_4, S_2', and S_1' are switched ON in order to attain an output level of $-(V_{DC,1} + V_{DC,2})$. Under such conditions, the antiparallel diode of S_2 gets a forward-biasing potential difference equal to $V_{DC,2}$, thereby short-circuiting the source $V_{DC,2}$. Therefore, at both positions S_2 and S_3, bidirectional-conducting-bidirectional-blocking switches are to be used.

Three such options are shown in Fig. 5.7. The switch shown in Fig. 5.7A is easy to control, however its ON state voltage drop will be high, as it is equal to the sum of voltage drops of two diodes and the IGBT. The switch shown in Fig. 5.7B has low ON state drop but needs two IGBTs and also requires two isolated gate driver circuits. A common emitter combination is shown in Fig. 5.7C having similar conduction losses to that of Fig. 5.7B, but requires only one gate driver circuit. This switch configuration has been chosen for Switches S_2 and S_3' of the BCS-MLI structure with two input DC sources as shown in Fig. 5.8.

5.3 SIMULATION STUDIES

To examine the performance of the BCS-MLI topology, a simulation model of a single-phase nine-level inverter is developed in the MATLAB/Simulink environment based on Fig. 5.8. Two input DC sources with $V_{DC,1} = 36$ V and $V_{DC,2} = 12$ V are used to obtain a trinary arrangement. The inverter is modulated using triangular carriers of frequency 1 kHz each disposed in phase-opposition-disposition and a sinusoidal reference of 50 Hz frequency with an amplitude modulation index of 0.85. Gate pulses for various switches are shown in Fig. 5.9.

Output voltage waveform and load current waveforms are shown in Fig. 5.10. An arbitrarily chosen R–L load with $R = 30\ \Omega$ and $L = 3.25$ mH is considered initially and changed to $R = 14\ \Omega$ and $L = 3.25$ mH at $t = 0.06$ s

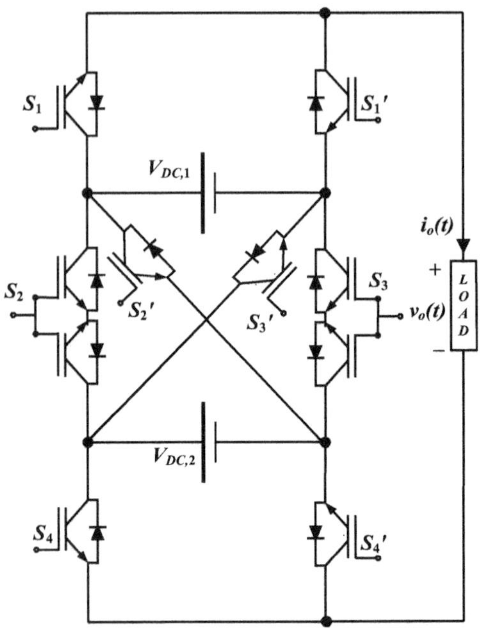

FIGURE 5.8 BCS-MLI with two input sources.

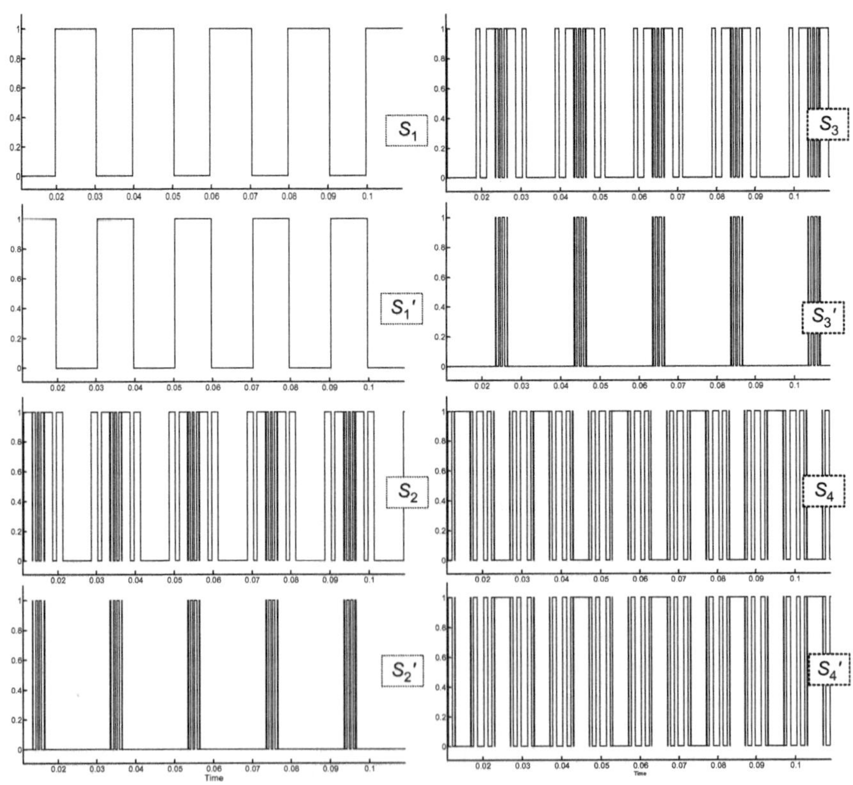

FIGURE 5.9 Gate pulses for various switches in a BCS-MLI topology-based nine-level inverter.

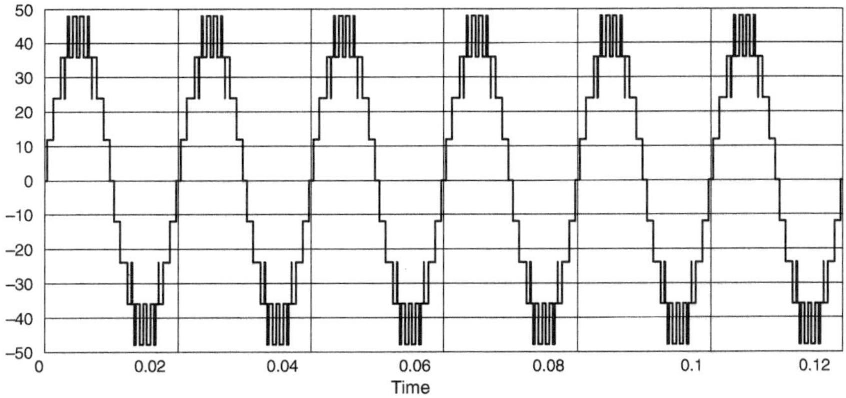

(A)

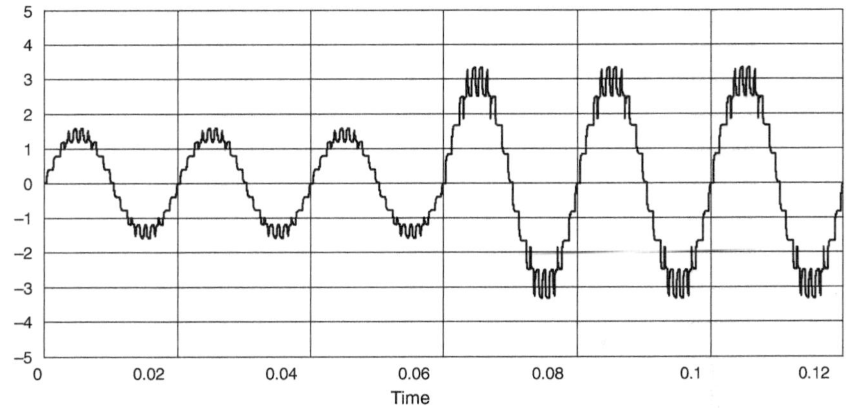

(B)

FIGURE 5.10 (A) Simulated nine-level output voltage waveform for a BCS-MLI topology-based inverter. (B) Current waveform for an RL load with increment of load at $t = 0.06$ s.

to observe the transients. It is observed that the output voltage is a nine-level waveform in equal steps of 12 V each. Also, the voltage waveform remains unperturbed when the load is increased. The current waveform indicates the inductive nature of the load.

Harmonic profiles of voltage and current waveforms are shown in Fig. 5.11. While the nine-level voltage has a THD of 15.33%, the initial load current shows a THD of 10.87% ($t < 0.06$ s) and final current (i.e., for $t \geq 0.06$ s) shows a THD of 12.26%. These harmonic profiles can be improved by proper selection of switching angles and/or employing filters but it is beyond the scope of this discussion as the treatment is restricted to description of the topology and issues closely related to it.

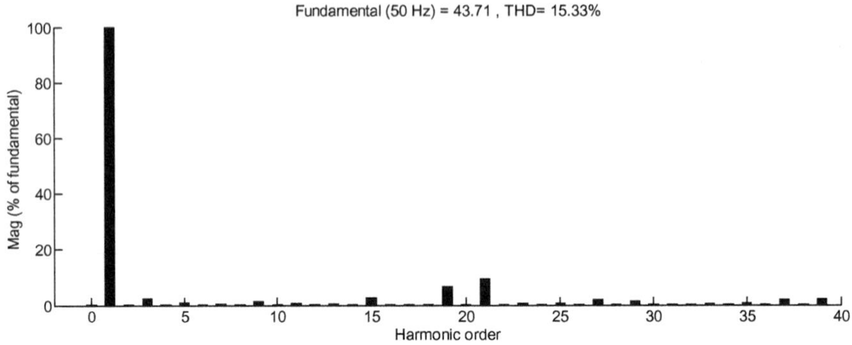

(A)

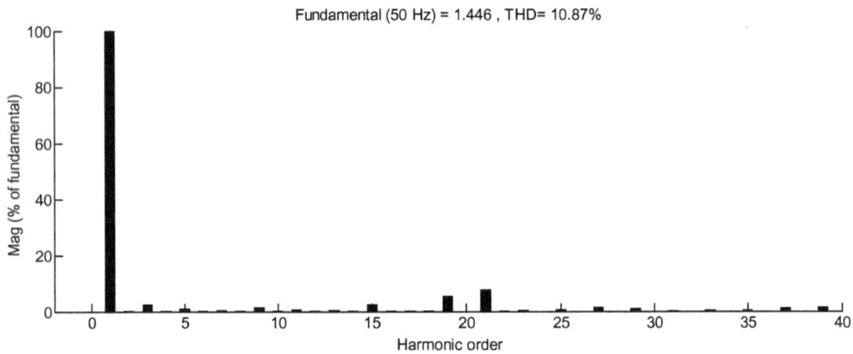

(B)

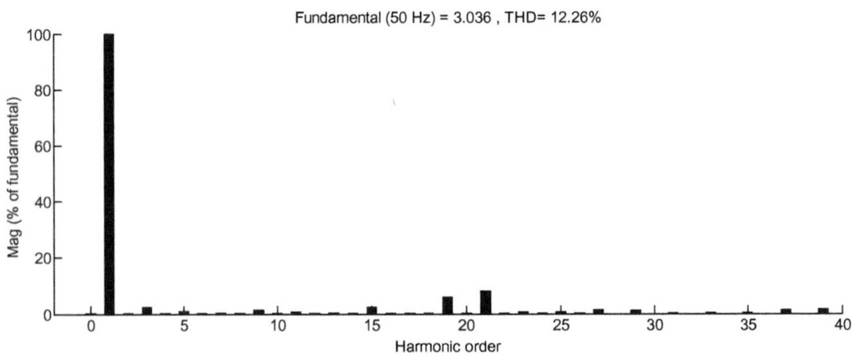

(C)

FIGURE 5.11 (A) Harmonic spectrum of output voltage of nine-level BCSMLI. (B) Harmonic spectrum of current waveform ($t < 0.06$ s). (C) Harmonic spectrum of current waveform ($t \geq 0.06$ s).

5.4 EXPERIMENTAL VALIDATION OF NINE-LEVEL BCS-MLI WITH TRINARY SOURCES

To validate the BCS-MLI topology, a prototype of a single-phase nine-level inverter has been tested in the laboratory. MOSFETs (IRF460) powered with suitable gate drivers are used as switching devices. Two batteries with voltages $V_{DC,1} = 36$ V and $V_{DC,2} = 12$ V are used as DC sources. A dSPACE DS1103 real-time controller interfaced with the MATLAB/Simulink model is used to generate the switching pulses shown in Fig. 5.12.

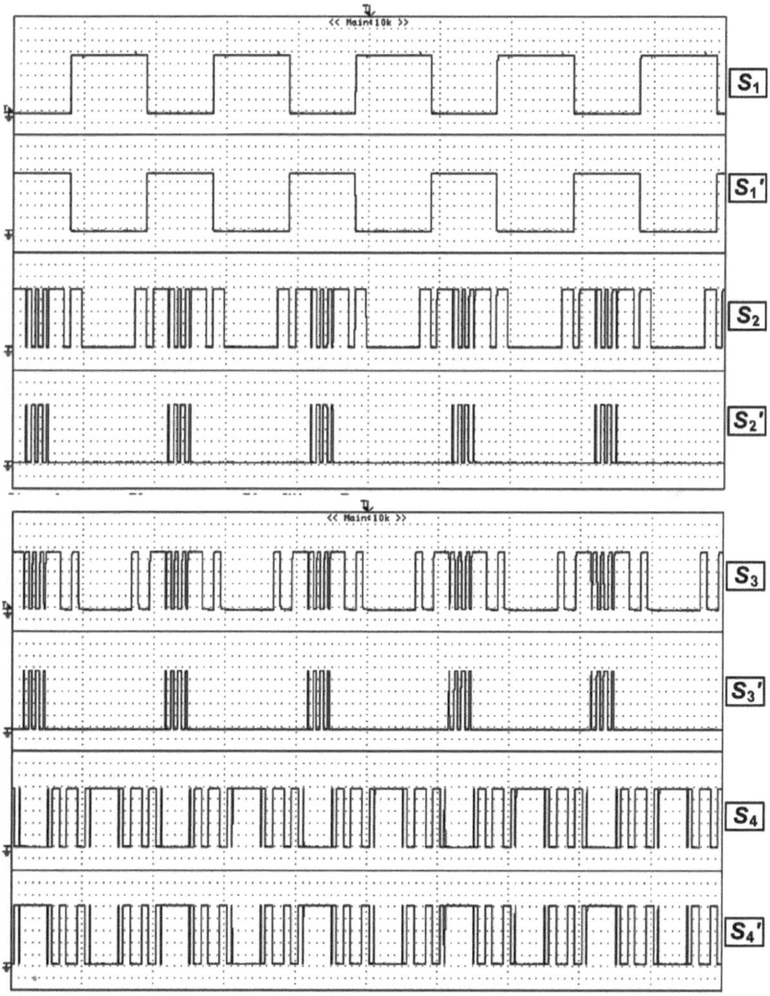

FIGURE 5.12 Gate pulses obtained from dSPACE DS1103 for experimental set-up of a nine--level BCS-MLI (all channels: scale on y-axis is 1 V/div).

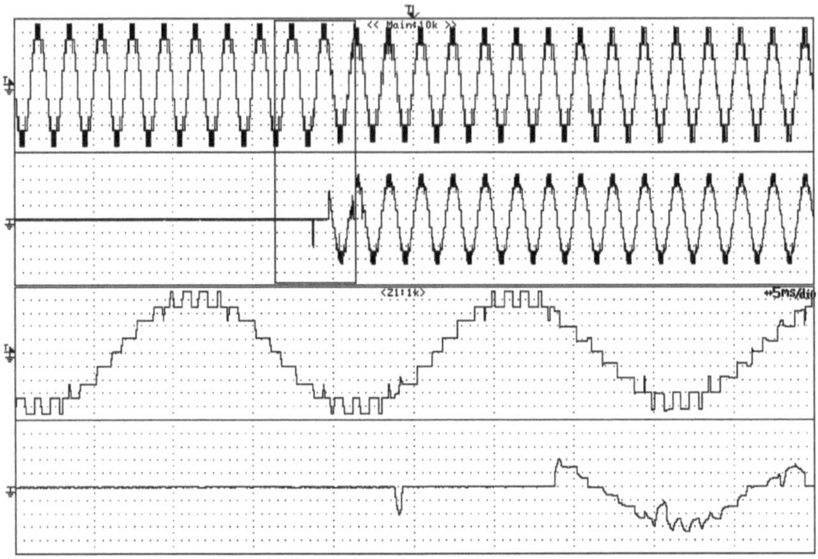

FIGURE 5.13 Switch-in response of the experimental BCS-MLI; voltage and current waveforms and zoomed-in views (y-axis : 10 V/div and 1 A/div, respectively).

The inverter is initially loaded with an inductive load of $R = 30\ \Omega$ and $L = 3.25$ mH which is subsequently changed to $R = 14\ \Omega$ and $L = 3.25$ mH to observe the transients. The output voltage and current waveforms for the switch-in response are shown in Fig. 5.13. The voltage comprises nine levels in equal steps of 12 V as expected. Fig. 5.14 shows the response of the inverter when the load is increased. It can be observed that the voltage remains expectedly unaffected with change in load. The harmonic spectrum of these waveforms is shown in Fig. 5.15. While the voltage shows a THD of 14.4%, the initial load current has a THD of 11.0%. After the load is changed, the load current exhibits a THD of 12.7%. The voltage waveform and its THD are in close agreement with the respective simulation results.

Similarly, the load currents have identical values for simulation and experimental studies with respect to their THD values and harmonic profiles.

5.5 CHARGE BALANCE CONTROL IN ASYMMETRICALLY CONFIGURED BCS-MLI

The BCS-MLI topology utilizes multiple DC sources to synthesize a stepped waveform. The power drawn from the different voltage sources in an inverter varies as a function of modulation index, output voltage levels, and load power factor [16]. This may result in unsteady and, at times, unstable DC

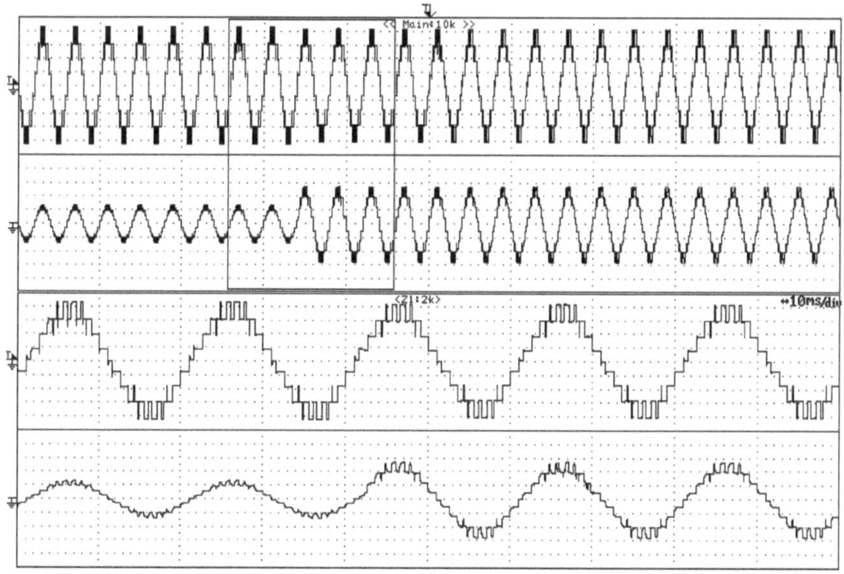

FIGURE 5.14 Voltage and current waveforms for the experimental BCS-MLI and respective zoomed-in views (y-axis : 10 V/div and 1 A/div, respectively).

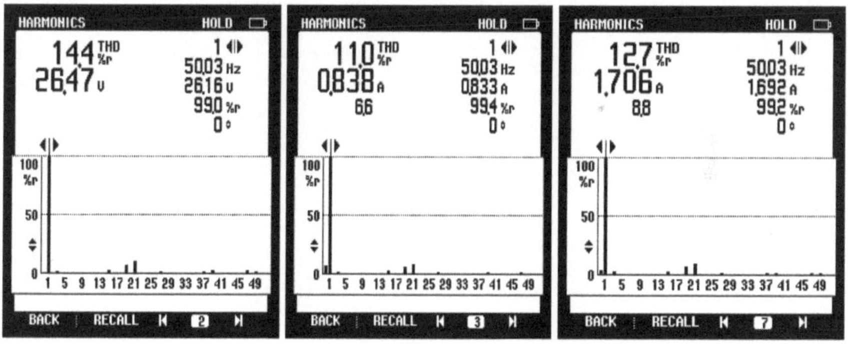

(A) (B) (C)

FIGURE 5.15 (A) Harmonic spectrum of the nine-level output voltage. (B) Harmonic spectrum of load current (for $R = 30\ \Omega$ and $L = 3.25$ mH). (C) Harmonic spectrum of load current (for $R = 14\ \Omega$ and $L = 3.25$ mH).

voltage levels. Hence the input DC sources deliver unequal power. As a result, different DC sources would have different lifetimes. Therefore, it is desirable that the input DC sources can manifest equal load sharing.

This section addresses the selection of an optimal asymmetric configuration of BCS-MLI with the possibility of charge balance control. Power balancing amongst the input DC sources in asymmetric multilevel inverters depends on the duty cycle of each source, which contributes toward synthesis of a given level at the output. When multiple input DC sources are used, it is desirable that these sources maintain a balanced state of charge or, in other words, exhibit equal load sharing. Available literature indicates that though charge balancing is achieved with symmetrical converters [17], it is not possible for binary and trinary configurations [16]. Charge balance control can be achieved if there are two or more input DC sources with equal voltages.

Table 5.2 summarizes various source configurations and corresponding availability of redundant states, the number of levels in the output voltage, and the possibility of charge balance control.

In light of the above discussion, it can be argued that an optimal configuration would involve a trade-off between asymmetry and charge-balancing considerations. In References [7,16], various source configurations are discussed with a greater number of redundant states as compared to binary or trinary configurations, along with the possibility of charge balance control.

TABLE 5.2 Typical Source Configurations and Charge Balancing Possibilities

S. no	Source Configuration	Number of Switches Required	Number of Levels in the Output Voltage	Number of Redundant States	Possibility of Charge Balancing
1	Unary $V_{DC,j} = V_{DC}$ $\{j = 1 \text{ to } N\}$	$4N$	$2N + 1$	$2^{2N} - 2N - 1$	Yes
2	Binary $V_{DC,j} = 2^{j-1} V_{DC}$ $\{j = 1 \text{ to } N\}$	$4N$	$2^{N++} - 1$	$2^{2N} - 2^{N+1} + 1$	No
3	Trinary $V_{DC,j} = 3^{j-1} V_{DC}$ $\{j = 1 \text{ to } N\}$	$4N$	3^N	$2^{2N} - 3^N$	No
4	As discussed in Reference [16] $V_{DC,j} = 2V_{DC}$ $\{j = 1 \text{ to } N - 1\}$; and $V_{DC,N} = V_{DC}$	$4N$	$4N - 1$	$2^{2N} - 4N + 1$	Yes
5	As discussed in Reference [7] $V_{DC,j} = 3V_{DC}$ $\{j = 1 \text{ to } N - 1\}$; and $V_{DC,N} = V_{DC}$	$4N$	$6N - 3$	$2^{2N} - 4N + 1$	Yes

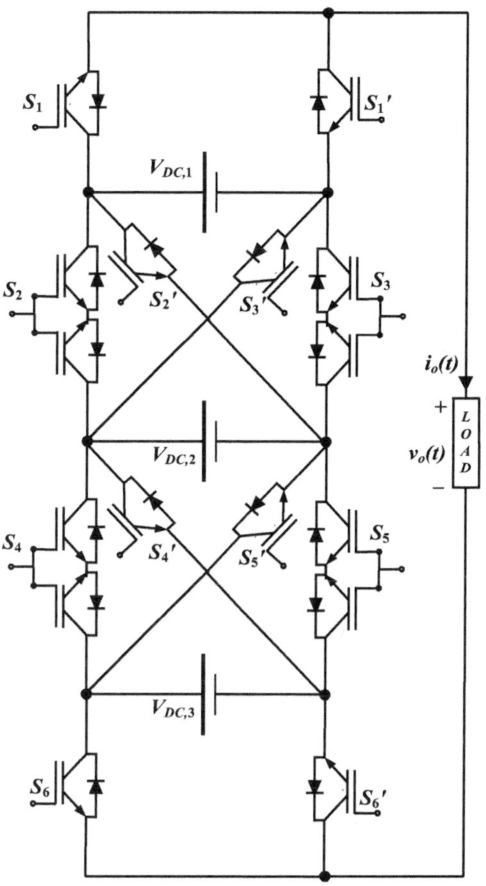

FIGURE 5.16 A 15-level single-phase inverter based on BCS-MLI topology with partially asymmetric source configuration.

Although BCS-MLI can be employed for all the configurations mentioned in Table 5.2, a source configuration shown in Fig. 5.16 is employed to demonstrate charge balance control with asymmetric sources. It has three input DC sources with voltages $V_{DC,1} = 36$ V, $V_{DC,2} = 36$ V, and $V_{DC,3} = 12$ V. Such a configuration is asymmetric and can synthesize 15 levels in the output waveform (-84 V to $+84$ V, in steps of 12 V each). Also, it has a total of $2^6 = 64$ states, of which 53 states would be redundant (number of redundant states $= 2^{2N} - 4N + 1$, $N =$ number of sources $= 3$). However, it should be emphasized that all voltage levels cannot have redundant states.

The fundamentals of the modulation scheme discussed in Chapter 4, Universal Control Scheme With Voltage-Level-Based Methods, are utilized to achieve charge balance control in the voltage sources $V_{DC,1}$ and $V_{DC,2}$. For a 15-level voltage waveform, the obtained aggregated signal *a(t)* will be as shown in Fig. 5.17A. Aggregated signal *a(t)* for a 15-level inverter is obtain with

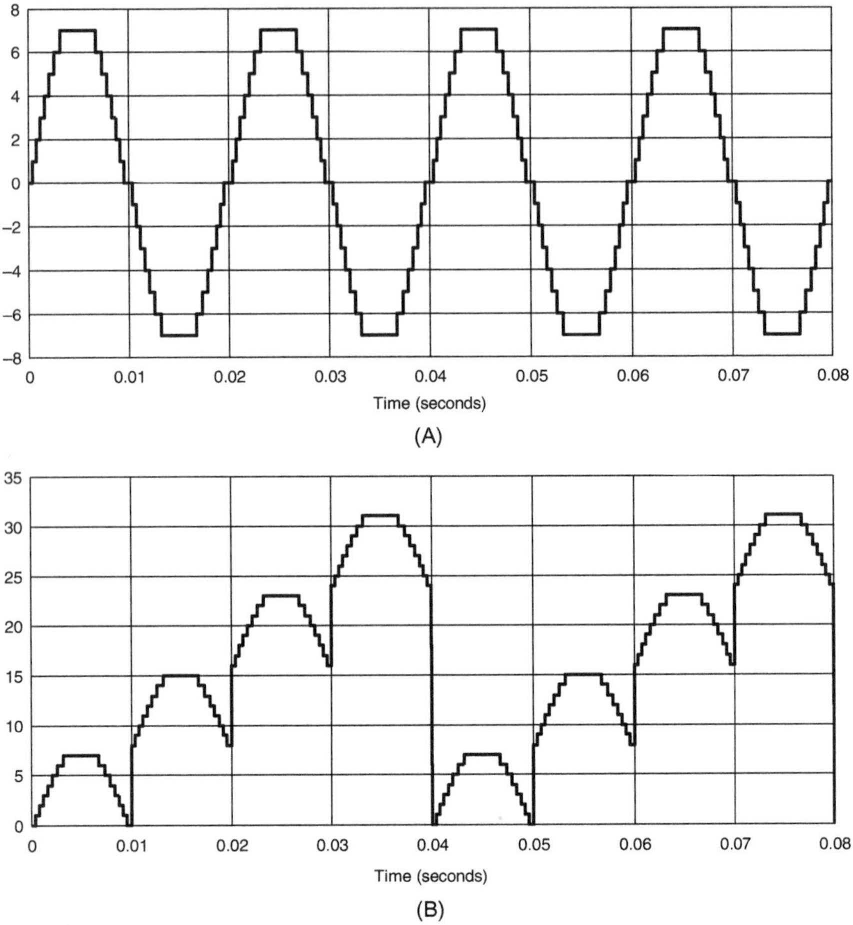

FIGURE 5.17 (A) Aggregated signal *a(t)* to obtain a 15-level output voltage. (B) Altered aggregated signal $a^{\#}(t)$ for a 15-level output waveform (equal load sharing amongst sources of the same voltage).

triangular carriers of frequency 100 Hz configured in phase-opposition-disposition and a sinusoidal reference of 50 Hz. Charge balance control can be achieved if sources $V_{DC,1}$ and $V_{DC,2}$ are utilized alternately in alternate cycles of the output waveform and thus balancing would be achieved in two cycles. Therefore, the aggregated signal "*a(t)*" is altered to "$a^{\#}(t)$" in a manner that its one cycle has four subparts as shown in Fig. 5.17B This signal has 33 levels and they are utilized as per the look-up table in Table 5.3.

A 15-level asymmetric multilevel inverter based on Fig. 5.16 is simulated in the MATLAB/Simulink environment using the SimPowerSystems toolbox. A highly inductive RL load with $R = 2\ \Omega$ and $L = 8$ mH is considered so as to observe the charge and discharge patterns of the DC sources. The output voltage

TABLE 5.3 Look-up Table for a 15-Level Inverter Based on BCS-MLI Topology for Equal Load Sharing Amongst Sources of Equal Voltages

Level of Altered Aggregated Signal $a^{\#}(t)$	Corresponding Voltage Output Level (V) ($V_{DC,1}=V_{DC,2}=36$ V, $V_{DC,3}=12$ V)	Switches in ON State (Other Switches Remain OFF)
0	0	S_1, S_2, S_4, S_6
1	$V_{DC,3}=12$	S_1, S_2, S_4, S'_6
2	$V_{DC,1} - V_{DC,3}=24$	S_1, S_3, S_5, S_6
3	$V_{DC,1}=36$	S_1, S_3, S_5, S'_6
4	$V_{DC,1}+V_{DC,3}=48$	S_1, S'_3, S_4, S'_6
5	$V_{DC,1}+V_{DC,2} - V_{DC,3}=60$	S_1, S'_3, S_5, S_6
6	$V_{DC,1}+ V_{DC,2}=72$	S_1, S'_3, S_5, S'_6
7	$V_{DC,1}+V_{DC,2}+V_{DC,3}=84$	S_1, S'_3, S'_5, S'_6
8	0	S_1, S_2, S_4, S_6
9	$-V_{DC,3}=-12$	S'_1, S_3, S_5, S_6
10	$V_{DC,3} - V_{DC,1}=-24$	S'_1, S_2, S_4, S'_6
11	$-V_{DC,1}=-36$	S'_1, S_2, S_4, S_6
12	$-V_{DC,1} - V_{DC,3}=-48$	S'_1, S'_2, S_5, S_6
13	$V_{DC,3} - V_{DC,1} - V_{DC,2}=-60$	S'_1, S'_2, S_4, S'_6
14	$-V_{DC,1} - V_{DC,2}=-72$	S'_1, S'_2, S_4, S_6
15	$-V_{DC,1} - V_{DC,2} - V_{DC,3}=-84$	S'_1, S'_2, S'_4, S_6
16	0	S_1, S_2, S_4, S_6
17	$V_{DC,3}=12$	S_1, S_2, S_4, S'_6
18	$V_{DC,2} - V_{DC,3}=24$	S_1, S_2, S'_5, S_6
19	$V_{DC,2}=36$	S_1, S_2, S'_5, S'_6
20	$V_{DC,2}+V_{DC,3}=48$	S_1, S_2, S_5, S'_6
21	$V_{DC,1}+V_{DC,2} - V_{DC,3}=60$	S_1, S_3, S'_5, S_6
22	$V_{DC,1}+V_{DC,2}=72$	S_1, S_3, S'_5, S'_6
23	$V_{DC,1}+V_{DC,2}+V_{DC,3}=84$	S_1, S_3, S_5, S'_6
24	0	S_1, S_2, S_4, S_6
25	$V_{DC,3}=-12$	S'_1, S'_3, S'_5, S_6
26	$V_{DC,3} - V_{DC,2}=-24$	S'_1, S'_3, S_4, S'_6
27	$-V_{DC,2}=-36$	S'_1, S'_3, S_4, S_6

(Continued)

TABLE 5.3 (Continued)

Level of Altered Aggregated Signal $a^{\#}(t)$	Corresponding Voltage Output Level (V) ($V_{DC,1} = V_{DC,2} = 36$ V, $V_{DC,3} = 12$ V)	Switches in ON State (Other Switches Remain OFF)
28	$-V_{DC,2} - V_{DC,3} = -48$	S'_1, S'_3, S'_4, S_6
29	$V_{DC,3} - V_{DC,1} - V_{DC,2} = -60$	S'_1, S'_2, S_4, S'_6
30	$-V_{DC,1} - V_{DC,2} = -72$	S'_1, S'_2, S'_4, S'_6
31	$-V_{DC,1} - V_{DC,2} - V_{DC,3} = -84$	S'_1, S'_2, S'_4, S_6

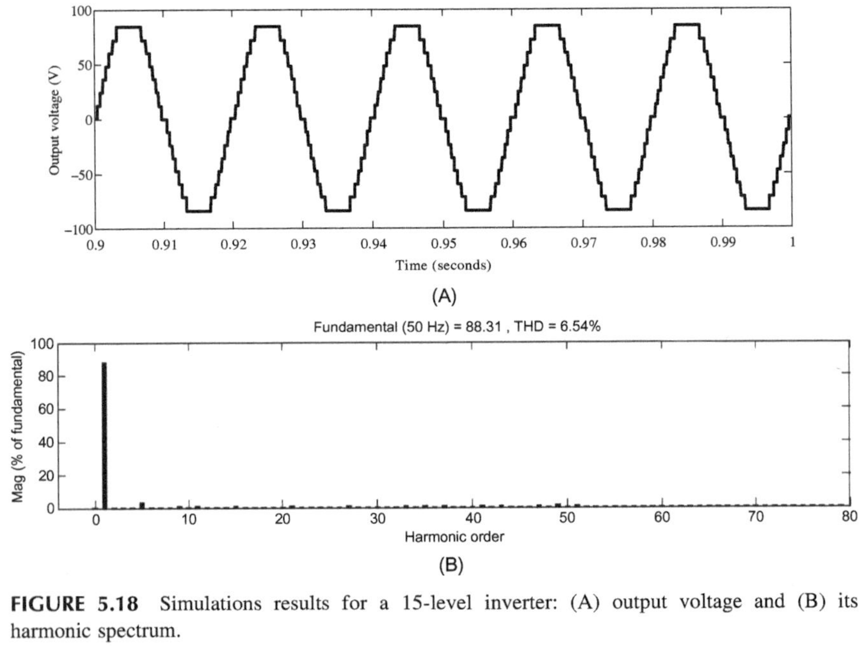

FIGURE 5.18 Simulations results for a 15-level inverter: (A) output voltage and (B) its harmonic spectrum.

waveform along with its harmonic spectrum is shown in Fig. 5.17. It is a 15-level stepped waveform in equal steps of 12 V. The current waveforms through the three DC sources are shown in Fig. 5.18. The current waveforms of two equal DC sources ($V_{DC,1}$ and $V_{DC,2}$) indicate an alternating pattern, which is a result of swapping the two sources for voltage level synthesis in alternate cycles. For example, the current waveform of source $V_{DC,1}$ for the duration $t = 0.94–0.96$ s is the same as that of source $V_{DC,2}$ for the duration $t = 0.96–0.98$ s and vice versa. Thus, the effective charge/discharge of these two sources is identical, thereby achieving equal load sharing between the two input sources.

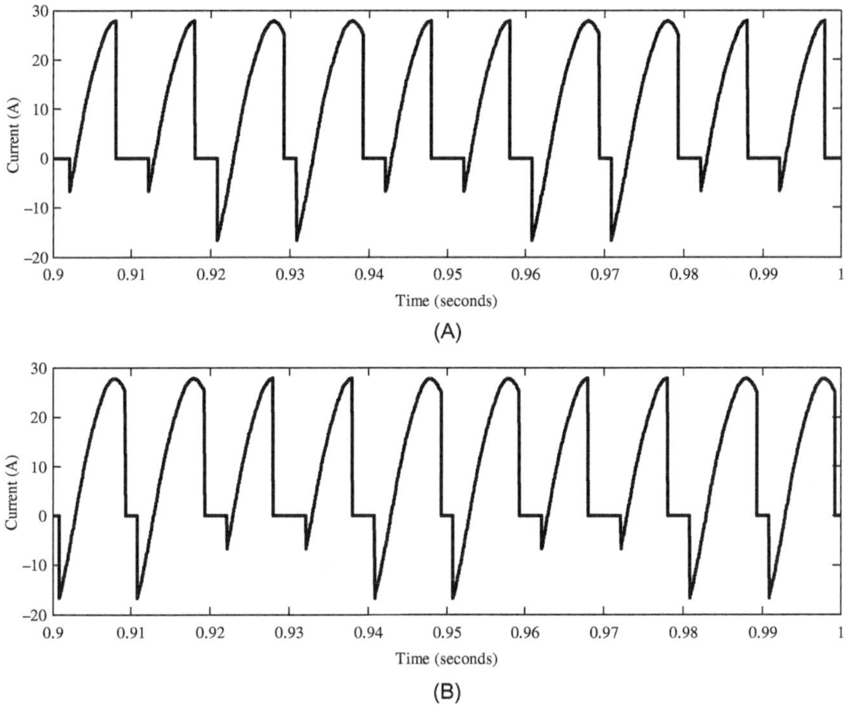

FIGURE 5.19 Current waveforms through DC sources in a 15-level inverter: (A) current through the source $V_{DC,1}$ and (B) current through the DC source $V_{DC,2}$.

TABLE 5.4 Power Measurements for Equal Sources in an Experimental 15-Level Inverter Based on BCS-MLI Topology

Output Power	Input Power		
	Power Delivered by Source $V_{DC,1}$	Power Delivered by Source $V_{DC,2}$	Power Delivered by Source $V_{DC,3}$
294.4 W	150.0 W	150.0 W	34.0 W

Experimental execution of the 15-level inverter using dSPACE DS1103 and batteries as DC sources is carried out to validate the scheme. A single-phase induction motor of 0.5 HP is used as the load. The output voltage waveform and the THD are shown in Fig. 5.19A and B, respectively. The output voltage waveform has equal steps of 12 V and a THD of 5%. Fig. 5.19A also shows the currents through voltage sources $V_{DC,1}$ and $V_{DC,2}$. The current waveforms reflect the effect of the power balancing scheme with similar alternate patterns in the wave shapes. The power measurements from the three sources are summarized in Table 5.4. It can be seen that both the sources $V_{DC,1}$ and $V_{DC,2}$

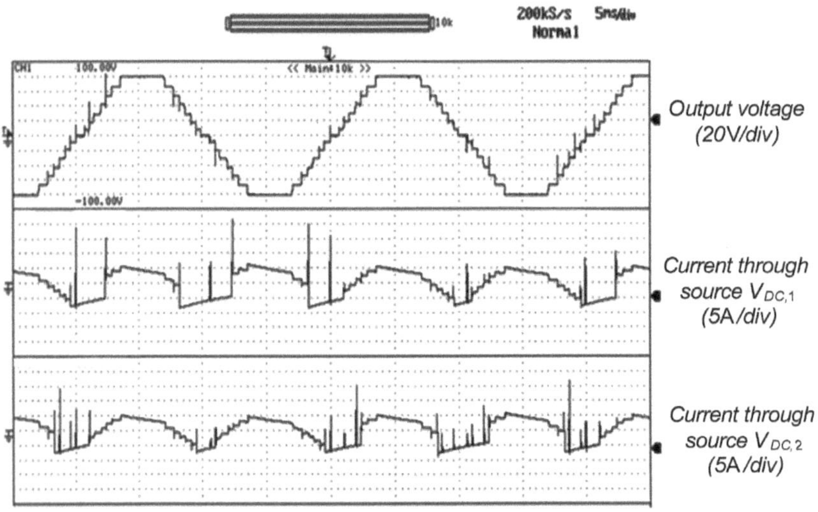

(A)

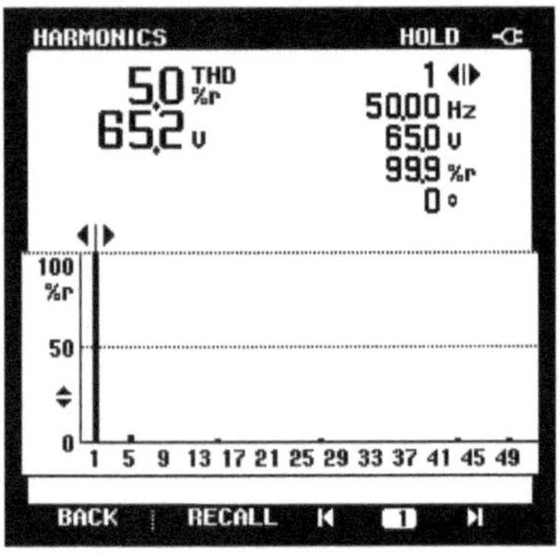

(B)

FIGURE 5.20 (A) Waveforms for experimental set-up of a 15-level inverter and (B) harmonic spectrum of a 15-level voltage.

make equal contributions, indicating an effective charge balancing control (Fig. 5.20).

5.6 CONCLUSION

Based on discussions in earlier chapters, this chapter discusses a wholesome description of topological innovations, with the example presentation of a new topology. The topology is named the BCS-MLI, and it can synthesize all possible additive and subtractive combinations of input DC levels in the output voltage waveform. This feature makes this topology eligible for all schemes of asymmetric source combinations. The input voltage levels are obtained from floating DC sources such that the consecutive sources are connected in a bridge fashion. The operation and performance of the proposed topology is ascertained through simulations and verified experimentally for a single-phase nine-level inverter. Moreover, a 15-level inverter with asymmetric source configuration is also investigated for charge balance control using the state selection procedure discussed in Chapter 4, Universal Control Scheme With Voltage-Level-Based Methods. The topology has two limitations: (1) it requires a mix of unidirectional-blocking-bidirectional-conducting and bidirectional-blocking-bidirectional-conducting switches and (2) asymmetric source configuration has to be compulsorily employed to extract complete mileage from the topology.

REFERENCES

[1] G.-J. Su, Multilevel DC-link inverter, IEEE Trans. Industry Appl. 41 (3) (2005) 848–854.

[2] E. Najafi, A.H.M. Yatim, Design and implementation of a new multilevel inverter topology, IEEE Trans. Ind. Electron. 59 (11) (2012) 4148–4154.

[3] S.-J. Park, F.-S. Kang, M.H. Lee, C.-U. Kim, A new single-phase five-level PWM inverter employing a deadbeat control scheme, IEEE Trans. Power Electron. 18 (3) (2003) 831–843.

[4] Z. Li, P. Wang, Y. Li, F. Gao, A novel single-phase five-level inverter with coupled inductors, IEEE Trans. Power Electron. 27 (6) (2012) 2716–2725.

[5] M.F. Kangarlu, E. Babaei, A generalized cascaded multilevel inverter using series connection of submultilevel inverters, IEEE Trans. Power Electron. 28 (2) (2013) 625–636.

[6] X. Kou, K.A. Corzine, Y.L. Familiant, "Full binary combination schema for floating voltage source multilevel inverters,", IEEE Trans. Power Electron. 17 (6) (2002) 891–897.

[7] S. Laali, K. Abbaszadeh, H. Lesani, A new algorithm to determine the magnitudes of dc voltage sources in asymmetric cascaded multilevel converters capable of using charge balance control methods, in: Electrical Machines and Systems (ICEMS), 2010 International Conference on, pp. 56–61, October 10–13, 2010.

[8] H. Sepahvand, J. Liao, M. Ferdowsi, Investigation on capacitor voltage regulation in cascaded h-bridge multilevel converters with fundamental frequency switching, IEEE Trans. Ind. Electron. 58 (11) (2011) 5102–5111.

[9] K.A. Corzine, X. Kou, Capacitor voltage balancing in full binary combination schema flying capacitor multilevel inverters, IEEE Pow. Electron. Lett. 1 (1) (2003) 2–5.

[10] J.I. Leon, L.G. Franquelo, R. Portillo, E. Dominguez, S. Vazquez, Improving multilevel flying capacitor converters features using new voltage ratios definitions, in: Electrotechnical Conference, 2006. MELECON 2006. IEEE Mediterranean, pp. 1190–1193, May 16–19, 2006.

[11] Y. Ounejjar, K. Al-Haddad, L.-A. Grégoire, Packed U cells multilevel converter topology: theoretical study and experimental validation, IEEE Trans. Ind. Electron. 58 (4) (2011) 1294–1306.

[12] G. Ceglia, V. Guzman, C. Sanchez, F. Ibanez, J. Walter, M.I. Gimenez, A new simplified multilevel inverter topology for DC -AC conversion, IEEE Trans. Power Electron. 21 (5) (2006) 1311–1319.

[13] Y. Hinago, H. Koizumi, A single phase multilevel inverter using switched series/parallel DC voltage sources, in: Energy Conversion Congress and Exposition, 2009. ECCE 2009. IEEE, pp. 1962–1967, September 20–24, 2009.
[14] M. Malinowski, K. Gopakumar, J. Rodriguez, M.A. Perez, A survey on cascaded multilevel inverters, IEEE Trans. Ind. Electron. 57 (7) (2010) 2197–2206.
[15] G. Mondal, K. Gopakumar, P.N. Tekwani, E. Levi, A reduced-switch-count five-level inverter with common-mode voltage elimination for an open-end winding induction motor drive, IEEE Trans. Ind. Electron. 54 (4) (2007) 2344–2351.
[16] E. Babaei, S.H. Hosseini, Charge balance control methods for asymmetrical cascade multilevel converters, in: Electrical Machines and Systems, 2007. ICEMS. International Conference on, pp. 74–79, October 8–11, 2007.
[17] C.-C. Hua, C.-W. Wu, C.-W. Chuang, A novel DC voltage charge balance control for cascaded inverters, IET Pow. Electron. 2 (2) (2009) 147–155.

Chapter 6

Cross-Connected Sources-Based Multilevel Inverter

6.1 INTRODUCTION

In Chapter 5, Multilevel Inverter Based on Bridge-Type Connected Sources, the so-called bridge-connected sources-based multilevel inverter (BCS-MLI) was presented. The BCS-MLI is capable of synthesizing all additive and subtractive combinations of input DC levels in the output waveform, and therefore, any pattern of asymmetric source configuration can be employed in it. It, however, has two important limitations:

1. It requires a mix of bidirectional-conducting-unidirectional-blocking and bidirectional-conducting-bidirectional-blocking power switches.
2. It requires asymmetric source configuration to maximize the number of output levels, so that the complexity in power switch requirements can be suitably justified.

While asymmetric input sources may not always be easily available, uniformity in power switch requirements can help in reducing assembly steps and economic manufacturing. Thus, it would be preferable to have a converter requiring similar power switches (i.e., all "unidirectional").

In this chapter, the topology described in Chapter 5, Multilevel Inverter Based on Bridge-Type Connected Sources, is modified to obtain another topology (named CCS-MLI), which can effectively operate with symmetric sources while requiring only bidirectional-conducting-unidirectional-blocking power switches. In addition, for symmetric source configuration, all switches, except four, are equally rated. These four switches are half-voltage rated as compared to the remaining ones. As a limitation, because the number of power switches is substantially reduced, the number of redundant states is reduced. Consequently, commonly popular asymmetric source configurations cannot be used in this topology. Hence, a new algorithm for configuring input sources asymmetrically in the topology is also presented in this chapter.

In addition, a detailed discussion is presented on how a two-source-based CCS-MLI has distinct advantages over the generalized structure: fundamental

Multilevel Inverters. DOI: https://doi.org/10.1016/B978-0-12-812448-2.00006-4

frequency of higher voltage-rated switches and equal load sharing among the input sources. Thus, the concept of "cascading" the multilevel cells is introduced. Once again, with the examples of these two topologies, the purpose is to understand the systematic development and analysis of novel topologies.

6.2 CCS-MLI TOPOLOGY AND PRINCIPLE OF OPERATION

A generalized single-phase structure of the topology is shown in Fig. 6.1. It has N number of electrically isolated input DC sources, which can be symmetrical or asymmetrical. The consecutive input DC sources are cross-connected via power switches, i.e., the higher potential terminal of the preceding source is connected to the lower potential terminal of the succeeding source and vice versa. The topology, henceforth, is referred to as "cross-connected sources-based multilevel inverter (CCS-MLI)." It can be observed that this structure can be simply obtained by removing all the bidirectional power switches from the BCS-MLI, as discussed in Chapter 5, Multilevel Inverter Based on Bridge-Type Connected Sources.

Input sources are designated as $V_{DC,j}$ {where, $j = 1$ to N}. Instantaneous current drawn from each source is designated as $i_j(t)$ {where, $j = 1$ to N}. Semiconductor switches in the power circuit are illustrated as IGBTs with antiparallel diodes and complementary switch pairs are designated as (S_j, S_j') {where, $j = 1$ to $N + 1$}. Load voltage and load currents are designated as $v_o(t)$ and $i_o(t)$, respectively. Also, nodal voltages are shown as $\delta_j(t)$ {$j = 1$ to $N + 1$}. Switch currents are indicated with $i_{Sj}(t)$ {for $j = 1$ to $N + 1$}.

Working of the topology can be comprehended with the help of a single-phase inverter having two input DC sources, as shown in Fig. 6.2. Switch pairs (S_j, S_j') {$j = 1, 2, 3$} are complementary. Therefore, three independent

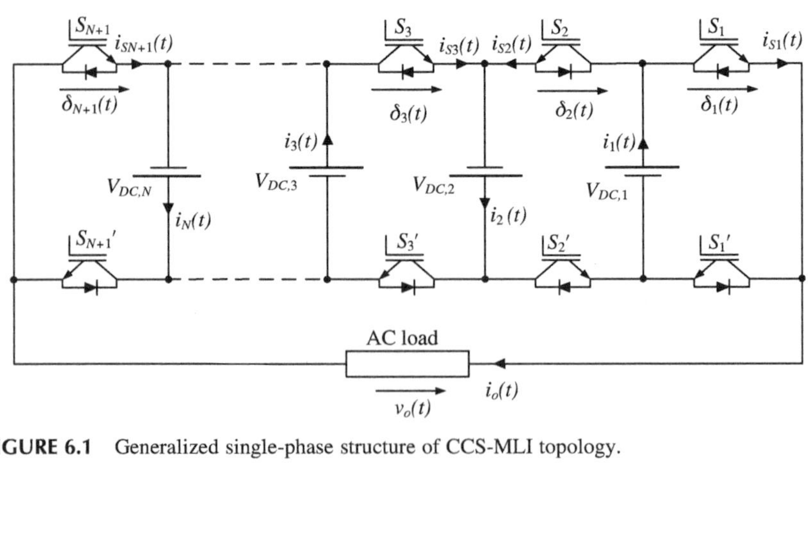

FIGURE 6.1 Generalized single-phase structure of CCS-MLI topology.

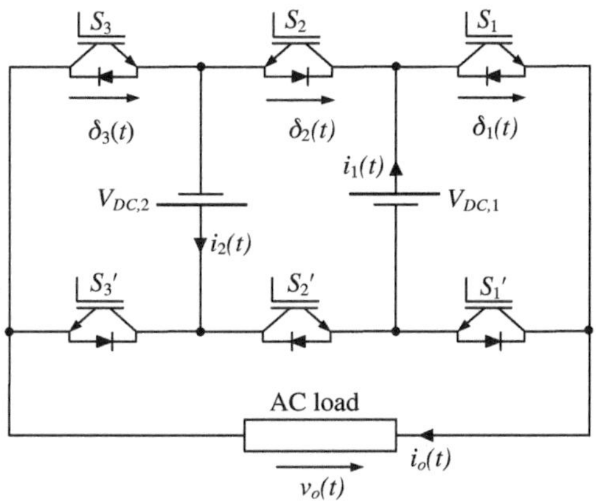

FIGURE 6.2 Single-phase structure of CCS-MLI topology with two input sources.

switches S_1, S_2, and S_3 would lead to the eight working states shown in Fig. 6.3.

Table 6.1 summarizes the states, switch positions, nodal voltages, and source currents. It is seen that to synthesize a given level, three switches conduct simultaneously. For example, to obtain a terminal voltage $v_o(t) = V_{DC,1} + V_{DC,2}$, switches S_1, S_2', and S_3 are turned ON.

A three-phase structure can also be realized using three single-phase legs. A three-phase structure based on CCS-MLI is illustrated in Fig. 6.4.

6.3 MATHEMATICAL FORMULATIONS FOR CCS-MLI

For a comprehensive understanding of a power converter, it is important to establish the relationship of voltages (especially across the load and power switches) with the switching function and input voltage levels. Similarly, it is important to establish the relationship of instantaneous currents drawn (especially through the input voltage sources and power switches) with the switching function and the load. In this manner a complete mathematical model can be obtained [1]. Also, appropriate voltage and current ratings for various components are essential for proper design and implementation of the converter. Important mathematical formulations are discussed here for the CCS-MLI topology.

6.3.1 Switching Function for Output Voltage

Associated with a given switch S, let a switching function $T(t)$ be defined such that,

$$T(t) = \begin{cases} 1, & \text{if switch } S \text{ is ON,} \\ 0, & \text{if switch } S \text{ is OFF.} \end{cases} \tag{6.1}$$

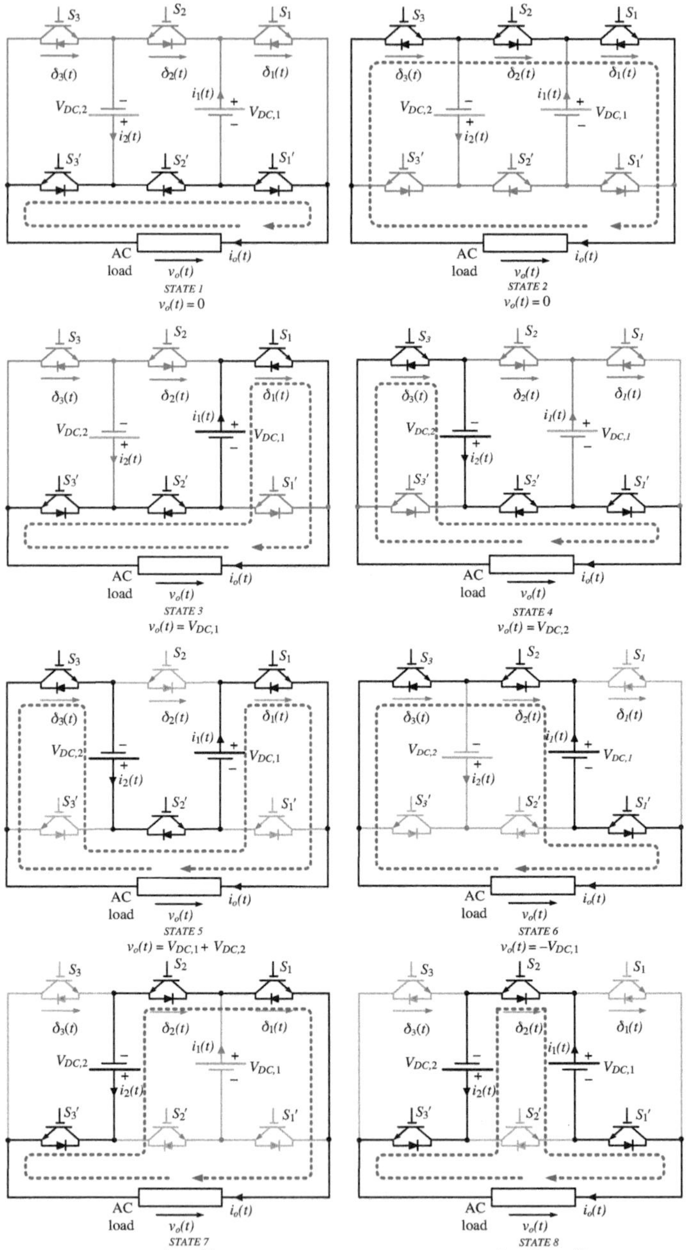

FIGURE 6.3 Working states for single-phase CCS-MLI with two input sources.

TABLE 6.1 Working States for CCS-MLI with Two Input Sources

State	Switch States (1 = ON, 0 = OFF)			Nodal Voltages			Source Currents		Output Voltage $v_o(t)$
	S_1	S_2	S_3	$\Delta_1(T)$	$\Delta_2(T)$	$\Delta_3(T)$	$I_1(T)$	$I_2(T)$	
1	0	0	0	$-V_{DC,1}$	$V_{DC,1}+V_{DC,2}$	$-V_{DC,2}$	0	0	0
2	1	1	1	0	0	0	0	0	0
3	1	0	0	0	$V_{DC,1}+V_{DC,2}$	$-V_{DC,2}$	$i_o(t)$	0	$V_{DC,1}$
4	0	0	1	$-V_{DC,1}$	$V_{DC,1}+V_{DC,2}$	0	0	$i_o(t)$	$V_{DC,2}$
5	1	0	1	0	$V_{DC,1}+V_{DC,2}$	0	$i_o(t)$	$i_o(t)$	$V_{DC,1}+V_{DC,2}$
6	0	1	1	$-V_{DC,1}$	0	0	$-i_o(t)$	0	$-V_{DC,1}$
7	1	1	0	0	0	$-V_{DC,2}$	0	$-i_o(t)$	$-V_{DC,2}$
8	0	1	0	$-V_{DC,1}$	0	$-V_{DC,2}$	$-i_o(t)$	$-i_o(t)$	$-(V_{DC,1}+V_{DC,2})$

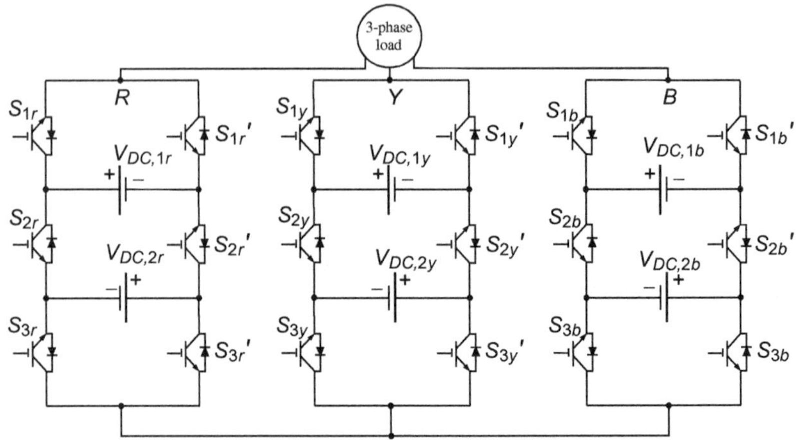

FIGURE 6.4 A three-phase inverter based on CCS-MLI topology.

Then, the load terminal voltage, $v_o(t)$, can be expressed in terms of nodal voltages $\delta_j(t)$ as,

$$v_o(t) = \sum_{j=1}^{N+1} \delta_j(t), \tag{6.2}$$

Such that,

$$\delta_j(t) = (-1)^j \{1 - T_j(t)\}(V_{\mathrm{DC},j} + V_{\mathrm{DC},j-1}), \tag{6.3}$$

Eqs. (6.2) and (6.3) imply

$$v_o(t) = \sum_{j=1}^{N+1} [\{(-1)^j (1 - T_j(t)\}(V_{\mathrm{DC},j} + V_{\mathrm{DC},j-1})] \tag{6.4}$$

Eq. (6.4) simplifies to

$$v_o(t) = \sum_{j=1}^{N+1} [\{T_j(t) - T_{j+1}(t)\}V_{\mathrm{DC},j}] \tag{6.5}$$

Here $T_{N+2}(t)$ should be taken as zero as there is no switch pair (S_{N+2}, S_{N+2}').

Eq. (6.5) is the governing equation for output voltage in terms of the values of input DC sources and switching functions.

6.3.2 Currents Through Input Sources

Source current $i_j(t)$ through a given source $V_{\mathrm{DC},j}$ $\{j = 1$ to $N\}$ can be expressed in terms of switching functions and load current $i_o(t)$ as,

$$i_j(t) = (-1)^{j+1} (T_j - T_{j+1}) i_0(t) \tag{6.6}$$

6.3.3 Currents Through Conducting Switches

Also, the instantaneous current in each switch is equal to the instantaneous load current when the switch is conducting and zero when the switch is blocking, i.e.,

$$i_{Sj}(t) = T_j(t).\ i_o(t) \tag{6.7}$$

6.3.4 Voltage Stress on Blocking Switches

Voltage stress $V_{\text{stress},j}$ borne by the switches of jth switch pair (S_j, S_j') can be expressed as:

$$V_{\text{stress},j} = V_{\text{DC},\,j-1} + V_{\text{DC},j} \tag{6.8}$$

where, $j = 1$ to $(N + 1)$, N being the number of input DC sources and $V_{\text{DC},0} = V_{\text{DC},N+1} = 0$.

It is extremely important to note here that the voltage stresses borne by the switches are different and are dependent on the source values. The voltage stress level has crucial consequences with respect to switching losses and switching frequency limitations [1], apart from costs. Further discussions on this issue are separately carried out for symmetrical and asymmetrical source configurations in the upcoming sections.

6.4 INVESTIGATIONS ON CCS-MLI WITH SYMMETRIC SOURCE CONFIGURATION

Source configuration for a multilevel topology is called symmetric when all the input sources have equal value [2], i.e., with reference to Fig. 6.1, $V_{\text{DC},1} = V_{\text{DC},2} = \ldots = V_{\text{DC},j} = \ldots = V_{\text{DC},N} = V_{\text{DC}}$. For N number of symmetric sources, all of them being equal to V_{DC}, the number of levels synthesized by the structure is given as

$$n_{\text{levels,symmetric}} = 2N + 1 \tag{6.9}$$

Also, the peak voltage attained for such a configuration is given by

$$v_{\text{max,symmetric}} = NV_{\text{DC}} \tag{6.10}$$

In this section, two investigations are carried out in relation to symmetric source configuration in CCS-MLI:

1. Fundamental frequency switching of the switches with higher voltage stress (i.e., the higher voltage-rated switches); and
2. Equal load sharing amongst the input DC sources.

6.4.1 Optimal Switching Operation of Five-Level Inverter Based on CCS-MLI

All the working states for CCS-MLI with two input sources are summarized in Table 6.1. Thus, with the use of these states, the load is fed by five levels, namely, $\pm 2V_{DC}$, $\pm V_{DC}$ and 0 for $V_{DC,1} = V_{DC,2} = V_{DC}$. With such a symmetric source configuration, states 3 and 4 become redundant for output level $+V_{DC}$ and states 6 and 7 become redundant for output level $-V_{DC}$. It is also important to mention here that for all the positive voltage levels and a zero level (states 1, 3, 4, and 5), switch S_2' always conducts, while for all the negative voltage levels and another zero level (states 2, 6, 7 and 8), switch S_2 always conducts. Therefore, it is possible to operate these two switches at fundamental frequency to obtain five output levels. This is a very significant feature because these two switches bear a voltage stress of $2V_{DC}$ each as compared to the rest of the switches which bear V_{DC} each.

To achieve fundamental frequency switching of high-voltage switches, states 1, 3 (or 4), and 5 need to be employed for synthesis of positive levels (including a zero level) and states 2, 6 (or 7), and 8 need to be employed for synthesis of negative voltage levels (including a zero level). This implies that utilization of six states is required for synthesis of five voltage levels. Hence the control scheme described in Chapter 4, Universal Control Scheme With Voltage-Level-Based Methods cannot be directly used as it uses five levels in the "aggregated signal" for a five-level output synthesis. Utilization of six states, however, would require six levels in the "aggregated signal."

A modified modulation scheme (with respect to the one presented in Chapter 4: Universal Control Scheme With Voltage-Level-Based Methods) is shown in Fig. 6.5A and the corresponding waveforms are shown in Fig. 6.5B. Four triangular waveforms of 1-kHz frequency each are used as carriers, configured in alternate phase opposition disposition (APOD) [3]. A sinusoidal waveform of 50-Hz frequency is taken as reference signal. Carrier signals above the zero reference are designated as $c_k^+(t)$ $\{k = 1, 2\}$ and those below the zero reference are designated as $c_k^-(t)$, $\{k = 1,2\}$. The reference is continuously compared with the carriers. If the reference is greater than carrier $c^+_k(t)$, the comparator gives "k" otherwise "$k-1$." If the reference is greater than carrier $c^-_k(t)$, the comparator gives "$-(k-1)$," otherwise "$-k$." The signals so obtained are then added so as to synthesize the "aggregated signal" $a(t)$, shown in Fig. 6.6. The switching pulses are derived from this signal by comparing it with the desired level and the output is fed to the switches corresponding to the level using the look-up table (shown in Table 6.2).

To examine the performance of the inverter and control scheme, a simulation model of a single-phase five-level inverter is developed in a MATLAB/Simulink environment. Two DC sources with $V_{DC,1} = V_{DC,2} = 24$ V are used. The switching pulses obtained are shown in Fig. 6.7, which shows that the switches S_2 and S_2' operate at a fundamental frequency of 50 Hz. Waveforms for load voltage and its harmonic spectrum are shown in Fig. 6.8. It is

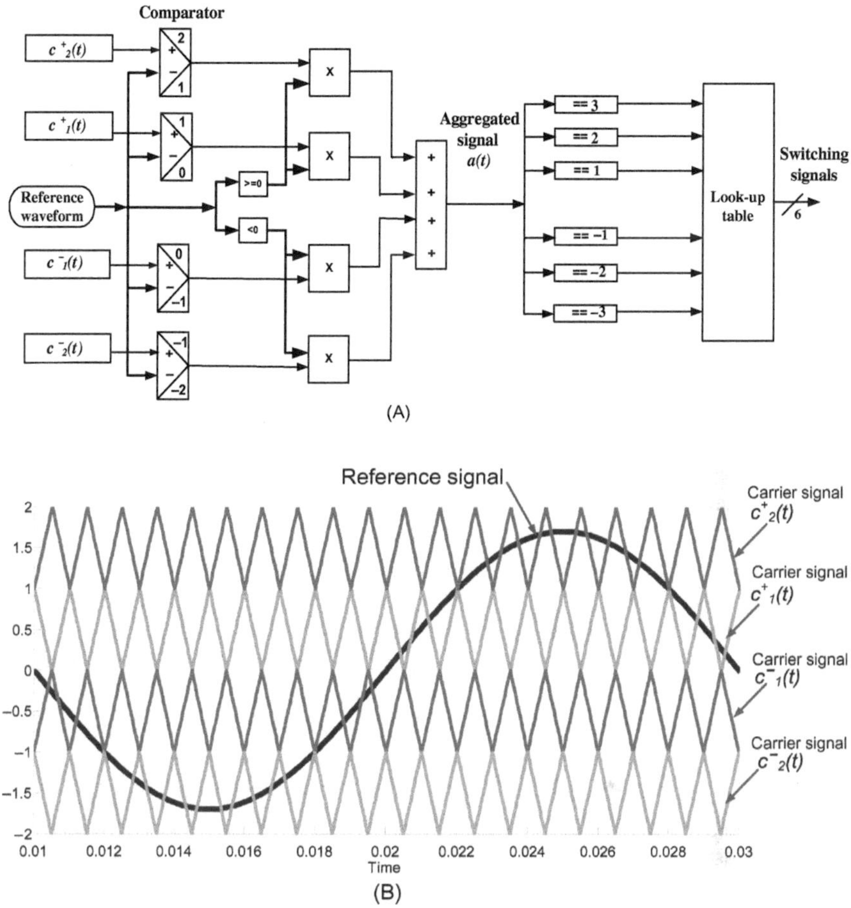

FIGURE 6.5 (A) Modulation scheme for a five-level inverter based on CCS-MLI topology. (B) Reference and carrier waveforms.

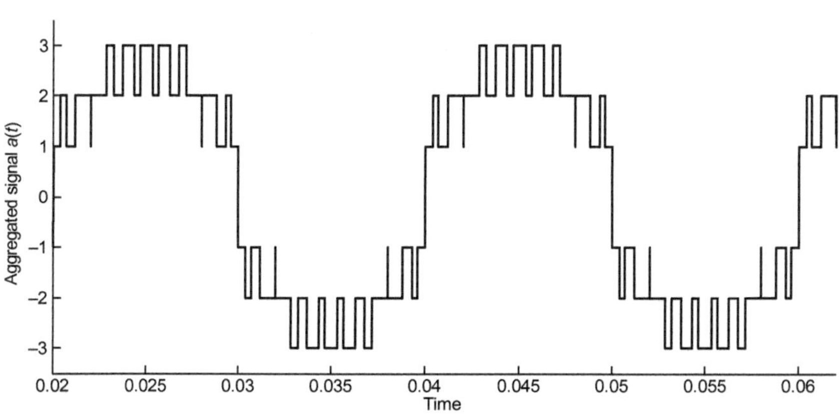

FIGURE 6.6 Aggregated signal $a(t)$ for a five-level inverter operated using six switching states.

TABLE 6.2 Look-Up Table for a Five-Level Inverter Based on CCS-MLI ($V_{DC,1} = V_{DC,2} = 24$ V)

Level of Aggregated Signal *a(t)*	State	Corresponding Output Voltage Level (V)	Switches in ON State
1	1	0	S'_1, S'_2, S'_3
2	3	24	S_1, S'_2, S'_3
3	5	48	S_1, S'_2, S_3
−1	2	0	S_1, S_2, S_3
−2	6	−24	S'_1, S_2, S_3
−3	8	−48	S'_1, S_2, S'_3

observed that the five-level voltage waveform has equal steps of 24 V each with a THD of 36.89%. Also, with an $R-L$ load ($R = 2\ \Omega$ and $L = 2$ mH), load current waveform and its harmonic spectrum are shown in Fig. 6.9A and B, respectively.

To validate the proposed topology and modulation strategy a prototype of single-phase five-level inverter has been developed in the laboratory. MOSFETs (IRF460) driven with suitable gate drivers are used as switching devices. Two batteries with voltages $V_{DC,\ 1} = V_{DC,2} = 24$ V are used as DC sources. The inverter is loaded with an inductive load with $R = 2\ \Omega$ and $L = 2$ mH. The control scheme is implemented through a dSPACE DS1103 real-time controller. The switching pulses obtained are shown in Fig. 6.10. The pulse patterns clearly indicate that switches S_2 and S'_2 operate at fundamental frequency. The load voltage obtained at the inverter terminals is shown in Fig. 6.11A, along with its harmonic spectrum shown in Fig. 6.11B. The load voltage is a stepped waveform with equal-sized steps of 24 V. The load current and its harmonic spectrum are shown in Fig. 6.12.

6.4.2 Equal Load Sharing Amongst the Input DC Sources in a Five-Level CCS-MLI

The CCS-MLI topology utilizes multiple DC sources to synthesize a stepped waveform. The power drawn from the different voltage sources in an inverter varies as a function of modulation index, output voltage levels, and load power factor. This may result in unsteady and, at times, unstable DC voltage levels. Hence, the input DC sources deliver unequal power. As a result, different DC sources would have different lifetimes. Therefore, it is desirable that the input DC sources can manifest equal load sharing. This subsection investigates the possibility of equal load sharing amongst the symmetrical

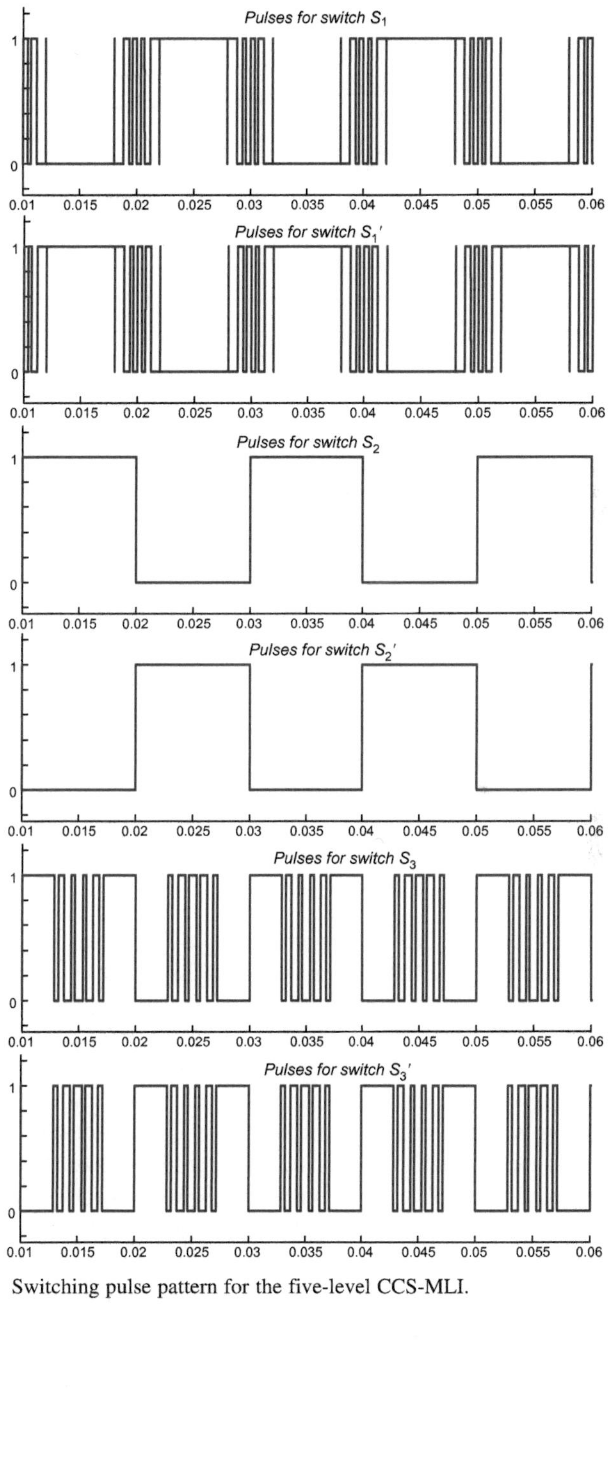

FIGURE 6.7 Switching pulse pattern for the five-level CCS-MLI.

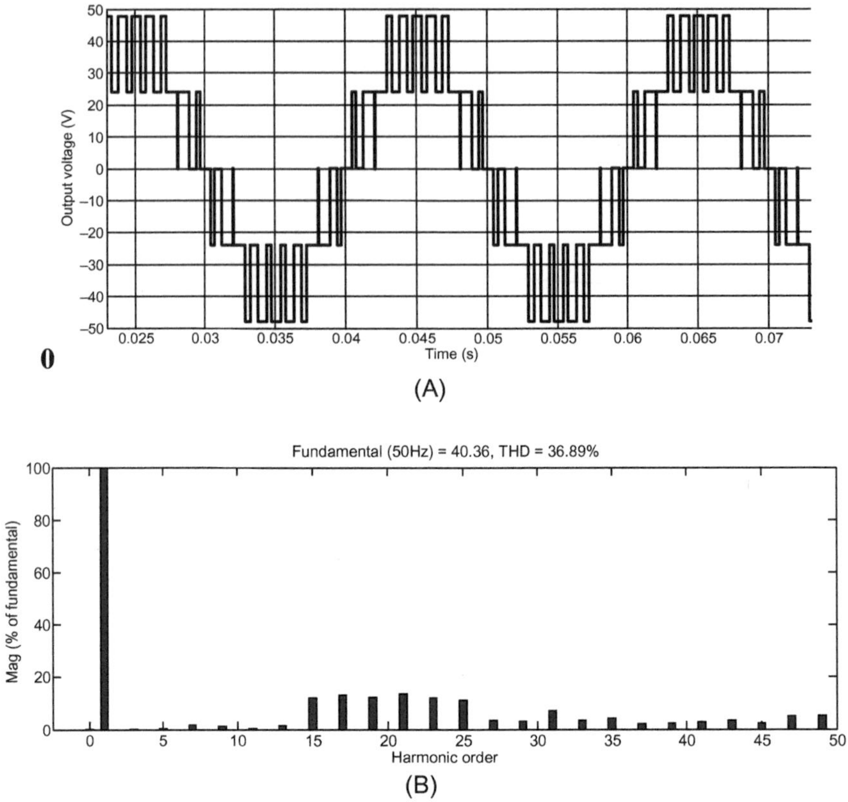

FIGURE 6.8 (A) Output voltage of the proposed five-level CCS-MLI structure. (B) Harmonic spectrum of the load voltage.

input DC sources for CCS-MLI. For simplicity, a five-level structure is analyzed first and then possibilities are explored for higher-level structures.

A five-level CCS-MLI as shown in Fig. 6.2 is assumed with input sources $V_{DC,1} = V_{DC,2} = V_{DC} = 24$ V. Equal load sharing can be achieved if sources $V_{DC,1}$ and $V_{DC,2}$ are utilized alternately in full cycles of the output waveform and thus balancing would be achieved in two cycles. For multilevel inverters, such load sharing is achieved using the so-called carrier-rotation scheme [4]. In order to implement equal load sharing, the control scheme previously discussed is employed for a five-level output with low switching frequency of high-voltage switches. In the scheme, the aggregated signal *a(t)* as shown in Fig. 6.6 is manipulated to obtain an altered aggregated signal *a*(t)* as indicated in Fig. 6.13, which basically consists of level-shifted alternate cycles of stepped waveform. The corresponding look-up table is shown as Table 6.3. It can be observed from the table that all the possible zero and nonzero states are utilized to achieve equal load sharing.

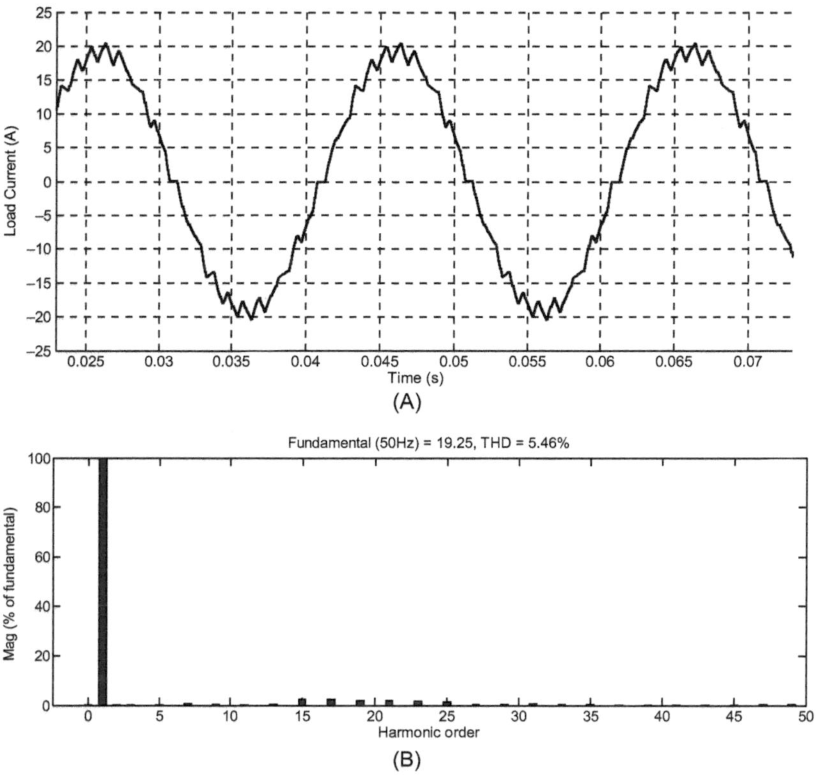

FIGURE 6.9 (A) Load current waveform of a five-level CCS-MLI with R–L load ($R = 2\ \Omega$ and $L = 2$ mH). (B) Harmonic spectrum of load current.

Based on the aforesaid scheme, a MATLAB/Simulink model is developed and the switching pulses obtained for the six switches are shown in Fig. 6.14.

Two important observations can be made from the switching waveforms:

- Switches S_2 and S_2' (both of which bear higher voltage stresses as compared to the rest of the switches) operate at a fundamental switching frequency of 50 Hz.
- Switches S_1 and S_3 have similar switching pulses with a phase difference of one time period of fundamental frequency (in this case the phase difference is of 0.02 seconds). This means that these two switches have equal utilization on average, with respect to switching performance. A similar argument can be made for switches S_1' and S_3'.

The aforesaid algorithm is validated experimentally for equal load sharing for a five-level inverter with $V_{DC,1} = V_{DC,2} = V_{DC} = 24$ V with an inductive load with ($R = 20\ \Omega$ and $L = 10$ mH). Experimental results are shown in Fig. 6.15,

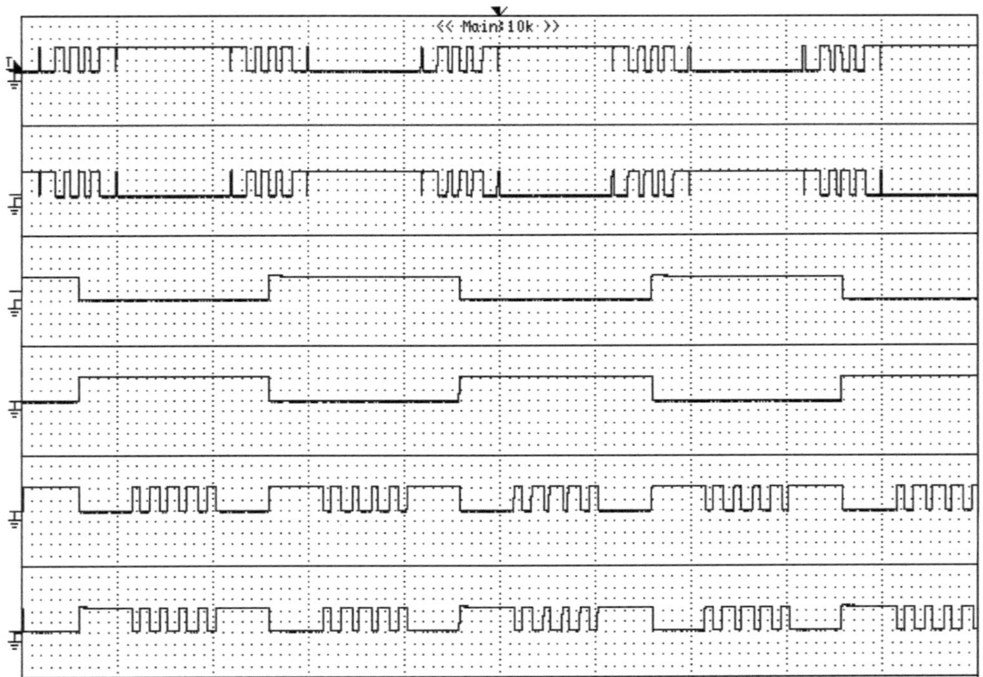

FIGURE 6.10 Experimentally obtained switching pulses for a five-level CCS-MLI (traces from top to bottom are for the switches S_1, S_1', S_2, S_2', S_3, and S_3', respectively).

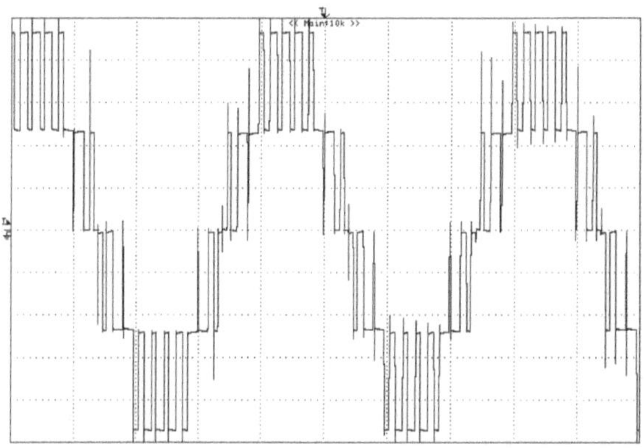

(A)

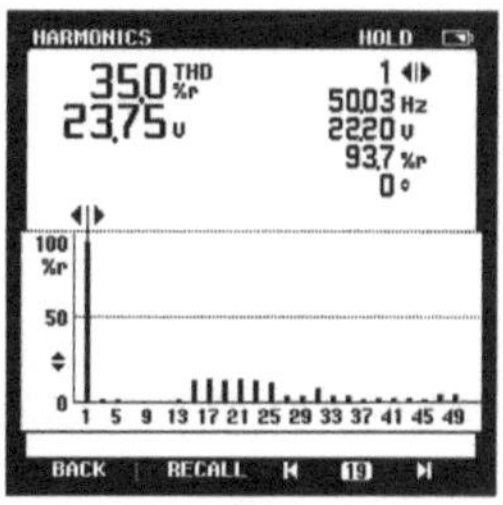

(B)

FIGURE 6.11 (A) Experimentally obtained output voltage for five-level CCS-MLI. (B) Harmonic spectrum of voltage.

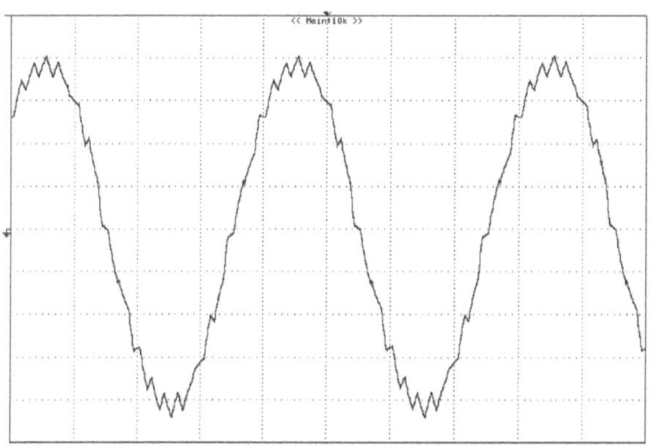

(A)

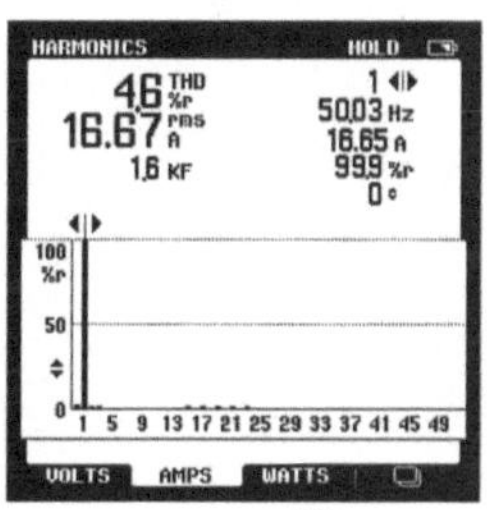

(B)

FIGURE 6.12 (A) Experimentally obtained load current waveform for five-level CCS-MLI (for $R = 2\ \Omega$ and $L = 2$ mH). (B) Harmonic spectrum of load current waveform.

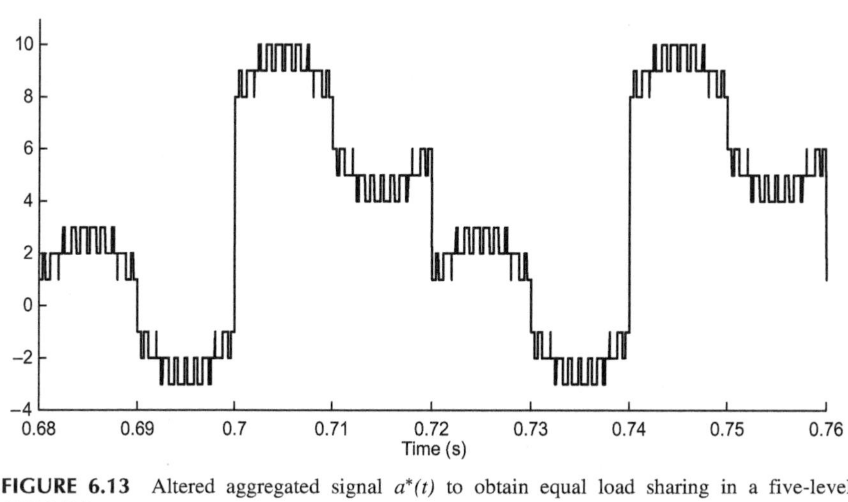

FIGURE 6.13 Altered aggregated signal $a^*(t)$ to obtain equal load sharing in a five-level CCS-MLI.

TABLE 6.3 Look-Up Table for a Five-Level Inverter Based on CCS-MLI for Equal Load Sharing Amongst Input DC Sources ($V_{DC,1} = V_{DC,2} = V_{DC} = 24$ V)

Level of Altered Aggregated Signal $a^*(t)$	State	Corresponding Voltage Output Level	Switches in ON State
−3	8	$-(V_{DC,1} + V_{DC,2})$	S_1', S_2, S_3'
−2	6	$-V_{DC,1}$	S_1', S_2, S_3
−1	2	0	S_1, S_2, S_3
1	1	0	S_1', S_2, S_3'
2	3	$V_{DC,1}$	S_1, S_2', S_3'
3	5	$(V_{DC,1} + V_{DC,2})$	S_1, S_2', S_3
4	8	$-(V_{DC,1} + V_{DC,2})$	S_1', S_2, S_3'
5	7	$-V_{DC,2}$	S_1, S_2, S_3'
6	2	0	S_1, S_2, S_3
8	1	0	S_1', S_2, S_3'
9	4	$V_{DC,2}$	S_1', S_2', S_3
10	5	$(V_{DC,1} + V_{DC,2})$	S_1, S_2', S_3

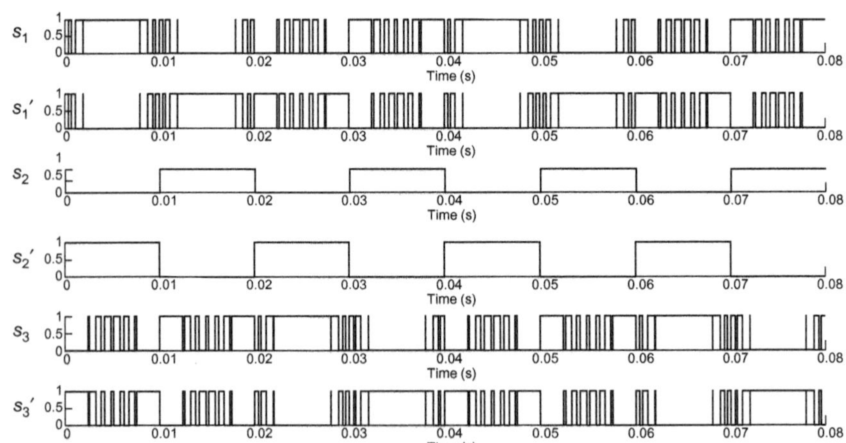

FIGURE 6.14 Switching pulse pattern for the a five-level CCS-MLI with equal load sharing.

in which the source currents indicate an alternating pattern which is a result of swapping the two sources for voltage level synthesis in alternate cycles. Thus, the effective charge/discharge of these two sources is identical, thereby achieving equal load sharing amongst the two sources. The power measurements

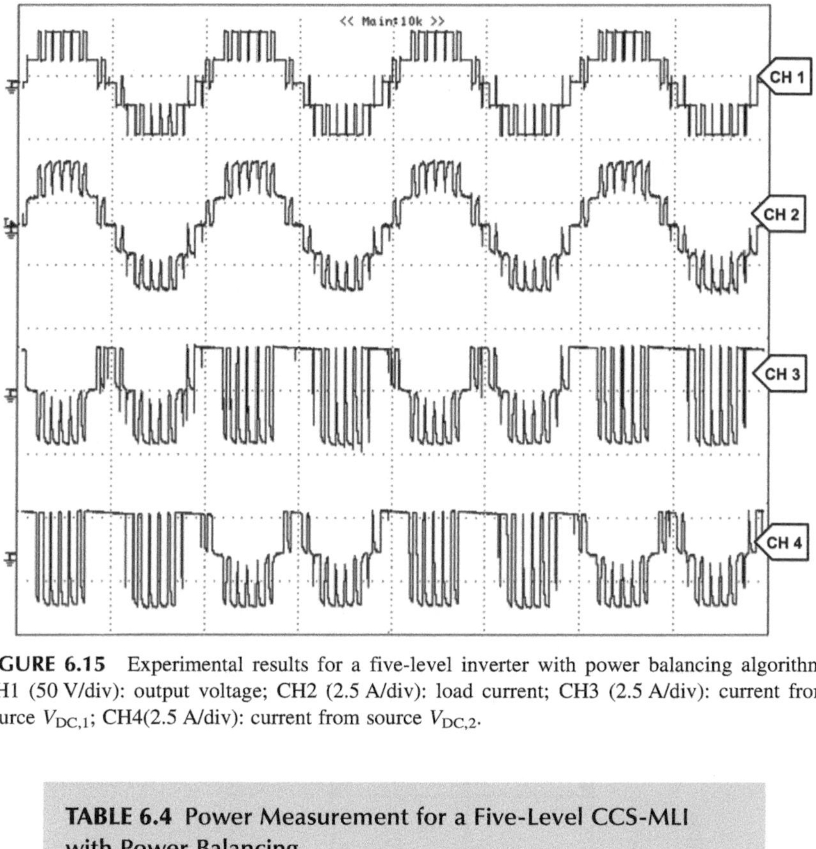

FIGURE 6.15 Experimental results for a five-level inverter with power balancing algorithm; CH1 (50 V/div): output voltage; CH2 (2.5 A/div): load current; CH3 (2.5 A/div): current from source $V_{DC,1}$; CH4(2.5 A/div): current from source $V_{DC,2}$.

TABLE 6.4 Power Measurement for a Five-Level CCS-MLI with Power Balancing

Output Power (W)	Input Power (W)	
	Source $V_{DC,1}$	Source $V_{DC,2}$
115	65	65

from input sources $V_{DC,1}$ and $V_{DC,2}$ are summarized in Table 6.4. It is seen that both sources make an equal contribution, indicating effective load sharing.

6.4.3 Optimal Switching Operation of Higher-Level Inverters (Number of Levels > 5) Based on CCS-MLI

In CCS-MLI with two input sources, fundamental switching of switches S_2 and S_2' is possible because switch S_2' conducts for all the positive voltage levels including zero and switch S_2 conducts for all the negative voltage levels including zero, as can be inferred from Table 6.1. In such a case it is possible to keep S_2' conducting for the positive half of fundamental switching and S_2 conducting for the negative half of fundamental switching, thereby synthesizing two positive and two negative levels, respectively, and zero level as well.

However, if a symmetric configuration with three input DC sources with $V_{DC,1} = V_{DC,2} = V_{DC,3} = V_{DC}$ is observed (shown in Fig. 6.16), it is seen that the higher voltage switches, namely, (S_2,S_2') and (S_3,S_3'), cannot possibly operate at fundamental switching frequency if a seven-level output is desired. This can be inferred from Table 6.5. For example, for a positive half cycle, if S_2' and S_3 conduct using states 2, 4, and 6, still to obtain zero level either switch S_2' has to commutated (to use state 7) or switch S_3 has to be commutated (to use state 8). Thus a strictly fundamental frequency operation is not possible for (S_2,S_2') and (S_3,S_3') if all seven levels (including 0) are desired at the output. A similar argument can also be made for higher-level structures.

However, the switching procedure as described in Section 6.4.1 can be utilized to minimize the switching frequencies of higher voltage switches. A simulation study is carried out using the aforesaid switching procedure and CCS-MLI

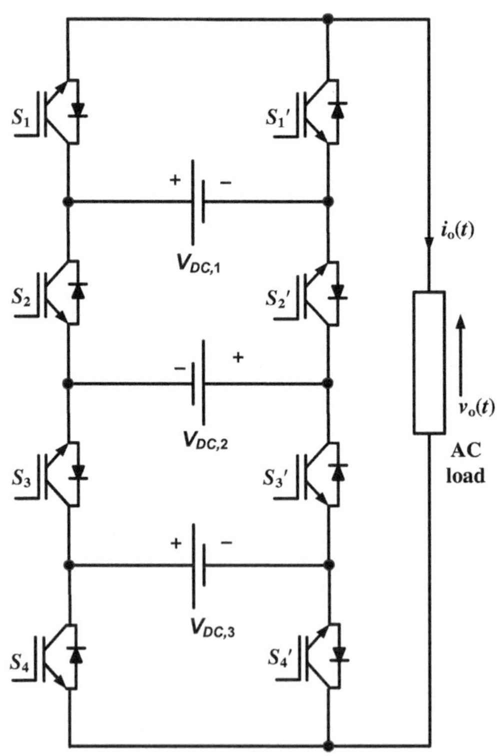

FIGURE 6.16 Single-phase CCS-MLI topology with three input sources.

TABLE 6.5 Switching Combinations for Positive Voltage Levels and Zero-Level Synthesis in CCS-MLI with Three Equal Input Sources ($V_{DC,1} = V_{DC,2} = V_{DC,3} = V_{DC}$)

State	Output Voltage Level	Source Combination	Conducting Switches
1	V_{DC}	$V_{DC,1}$	S_1, S_2', S_3', S_4'
2		$V_{DC,2}$	S_1', S_2', S_3, S_4
3		$V_{DC,3}$	S_1, S_2, S_3, S_4'
4	$2V_{DC}$	$V_{DC,1} + V_{DC,2}$	S_1, S_2', S_3, S_4
5		$V_{DC,2} + V_{DC,3}$	S_1, S_2, S_3, S_4
6	$3V_{DC}$	$V_{DC,1} + V_{DC,2} + V_{DC,3}$	S_1', S_2', S_3, S_4'
7	0	–	S_1, S_2, S_3, S_4
8		–	S_1', S_2', S_3', S_4'

with three input voltage sources are taken as $V_{DC,1} = V_{DC,2} = V_{DC,3} = V_{DC} = 100$ V and frequencies of carrier and reference waveforms are taken as 1 kHz and 50 Hz, respectively. Corresponding switching signals (for switches S_1, S_2, S_3, and S_4) are shown in Fig. 6.17, which shows that the higher voltage switches, namely, (S_2, S_2') and (S_3, S_3'), operate at much lower switching frequencies as compared to those of lower voltage level switches, namely, (S_1, S_1') and (S_4, S_4'). It can be observed that while switches S_1 and S_4 operate with a switching frequency of 1 kHz, switches S_2 and S_3 operate with a switching frequency of 300 Hz.

6.4.4 Equal Load Sharing in Higher-Level Inverters (Number of Levels > 5) Based on CCS-MLI

As discussed earlier, for equal load sharing amongst input DC sources, each source must contribute to the synthesis of levels alternately. If three DC sources, namely, $V_{DC,1} = V_{DC,2} = V_{DC,3} = V_{DC}$ are present, they should contribute towards synthesis of levels $\pm V_{DC}$, $\pm 2V_{DC}$, and $\pm 3V_{DC}$ in combinations of $V_{DC,1}/V_{DC,2}/V_{DC,3}$, $(V_{DC,1} + V_{DC,2})/(V_{DC,2} + V_{DC,3})/(V_{DC,3} + V_{DC,1})$ and $(V_{DC,1} + V_{DC,2} + V_{DC,3})$, respectively. However, it can be seen from Eq. (6.5) that no switching combination can synthesize the voltage levels $\pm (V_{DC,3} + V_{DC,1})$. Therefore, equal load sharing amongst the input DC sources cannot be achieved. A similar situation will occur where the number of input DC sources is greater than three. Hence, it can be concluded that equal load sharing amongst all the input sources cannot be achieved in the CCS-MLI when more than two sources are present. However, it is possible to attain

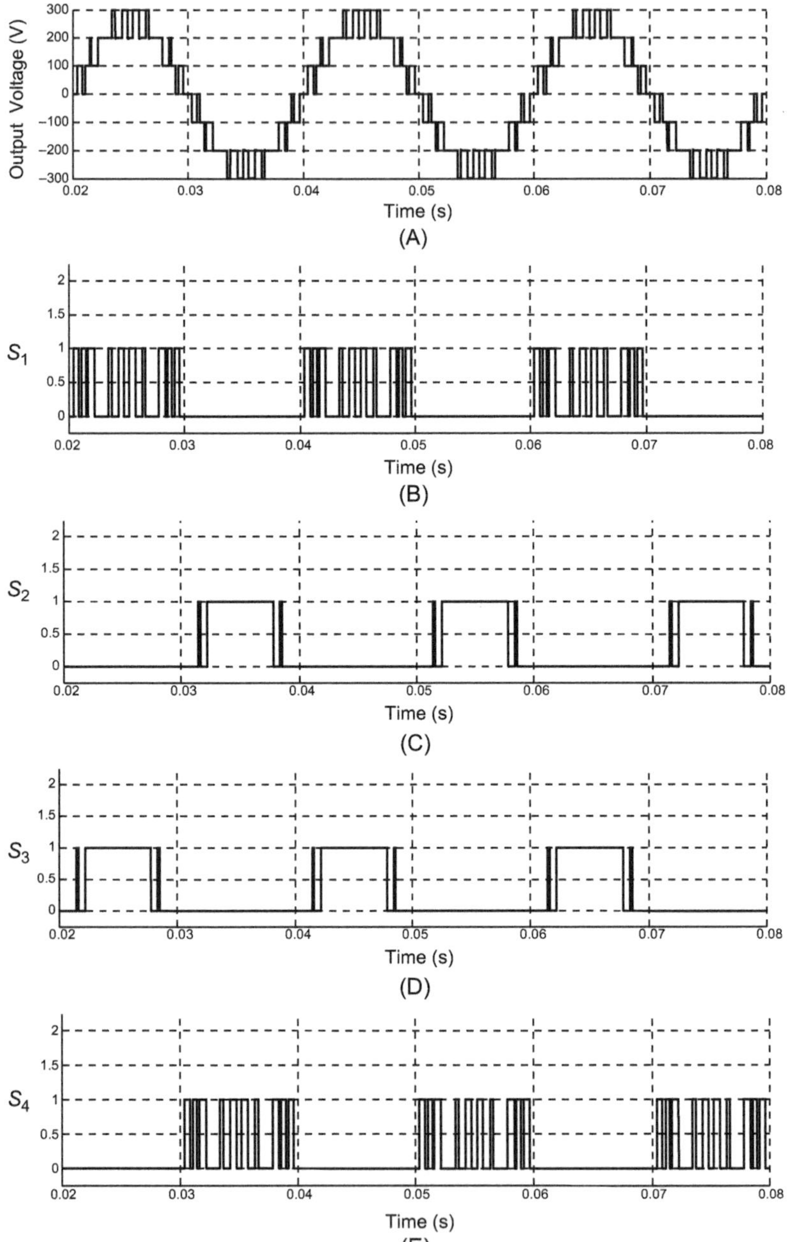

FIGURE 6.17 (A) Seven-level output for CCS-MLI using the proposed switching procedure; (B), (C), (D), and (E) are switching pulses for S_1, S_2, S_3, and S_4, respectively.

equal load sharing amongst some of the input sources. For example, again taking the case of three input sources, equal load sharing can be achieved amongst sources $V_{DC,1}$ and $V_{DC,2}$ or amongst sources $V_{DC,2}$ and $V_{DC,3}$. A similar treatment can be extended for structures with a higher number of input sources.

6.5 CCS-MLI WITH ASYMMETRIC SOURCE CONFIGURATION

When two or more of the input DC sources have unequal values, the configuration is called as an asymmetric source configuration. Such a configuration enables a lower number of power switches to be used for the same number of output levels as compared to a symmetric configuration [5].

In this section, the CCS-MLI is explored to synthesize a multilevel waveform with equal-sized steps, with an asymmetric source configuration. Two popular asymmetric source configurations employed in multilevel inverters are:

1. *Binary configuration*: This consists of DC sources having a geometric progression with a factor of "2," i.e.,

$$V_{DC,j} = 2^{j-1} V_{DC} \{j = 1 \text{ to } N\} \tag{6.11}$$

2. *Trinary configuration*: This consists of sources having a geometric progression with a factor of "3," i.e.,

$$V_{DC,j} = 3^{j-1} V_{DC} \{j = 1 \text{ to } N\} \tag{6.12}$$

For CCS-MLI, the asymmetric configurations discussed above cannot be employed satisfactorily. For example, if the CCS-MLI is simulated using a trinary source configuration with $V_{DC,1} = 100$ V and $V_{DC,2} = 300$ V, nine of the expected nine levels (−400 V to 400 V, in steps of 100 V each) are synthesized. For a nine-level inverter, carrier and reference signals are shown in Fig. 6.18A. A simulated output voltage waveform from CCS-MLI is shown in Fig. 6.18B which synthesizes seven levels with unequal step-sizes, whereas the desired nine-level waveform is shown in Fig. 6.18C.

Similarly, it can be shown that other commonly used asymmetric configurations cannot be implemented for the CCS-MLI. This is so because the structure does not synthesize all additive and subtractive combinations of the input DC levels. It can be inferred from Table 6.6 that there are some voltage levels which are "skipped," i.e., no switching combination exists which can possibly synthesize these levels; Eq. (6.5) helps to estimate which of the levels are present for possible switching combinations. Therefore, an appropriate algorithm for asymmetric source configuration is required which can synthesize a multilevel waveform with a maximum number of levels with equal step size.

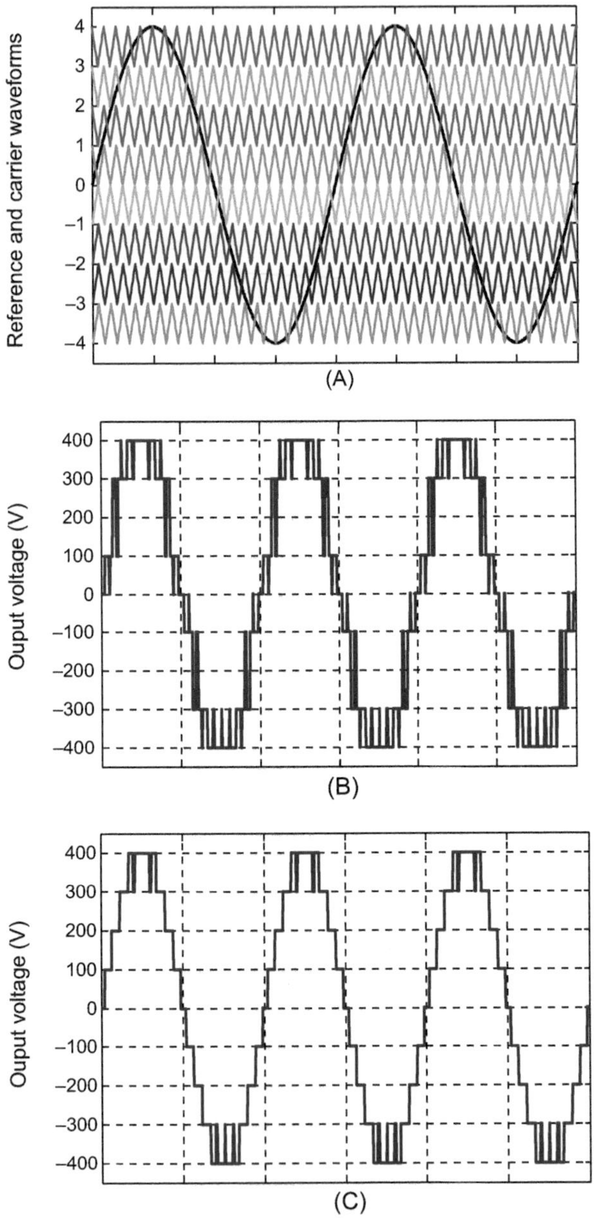

FIGURE 6.18 (A) Reference and carrier waveforms for a nine-level inverter. (B) Output voltage waveform for the CCS-MLI topology with trinary configuration ($V_{DC,1} = 100$ V and $V_{DC,2} = 300$ V). (C) Expected nine-level voltage waveform for an inverter with trinary configuration ($V_{DC,1} = 100$ V and $V_{DC,2} = 300$ V).

TABLE 6.6 Additive and Subtractive Combinations of Input DC Levels and Their Presence/Absence in CCS-MLI Topology

Number of Sources	Input DC Level(s)	Possible Voltage Levels in the Output and Possibility of Synthesis of the Level by CCS-MLI (in Brackets)	Total Number of Possible Combinations of Input DC Voltage Levels	Total Number of Possibilities Offered by the CCS-MLI Topology
One	$V_{DC,1}$	0 (Yes)	3	3
		$\pm (V_{DC,1})$ (Yes)		
Two	$V_{DC,1}$, $V_{DC,2}$	0 (Yes)	9	7
		$\pm V_{DC,1}$ (Yes)		
		$\pm V_{DC,2}$ (Yes)		
		$\pm (V_{DC,1} - V_{DC,2})$ (No)		
		$\pm (V_{DC,1} + V_{DC,2})$ (Yes)		
Three	$V_{DC,1}$, $V_{DC,2}$, $V_{DC,3}$	0 (Yes)	27	15
		$\pm V_{DC,1}$ (Yes)		
		$\pm V_{DC,2}$ (Yes)		
		$\pm V_{DC,3}$ (Yes)		
		$\pm (V_{DC,1} + V_{DC,2})$ (Yes)		
		$\pm (V_{DC,1} - V_{DC,2})$ (No)		
		$\pm (V_{DC,1} + V_{DC,3})$ (No)		
		$\pm (V_{DC,1} - V_{DC,3})$ (Yes)		
		$\pm (V_{DC,2} + V_{DC,3})$ (Yes)		
		$\pm (V_{DC,2} - V_{DC,3})$ (No)		
		$\pm (V_{DC,1} + V_{DC,2} + V_{DC,3})$ (Yes)		
		$\pm (V_{DC,1} + V_{Dc,2} - V_{DC,3})$ (No)		
		$\pm (V_{DC,1} - V_{DC,2} + V_{DC,3})$ (No)		
		$\pm (V_{DC,1} - V_{DC,2} - V_{DC,3})$ (No)		

6.5.1 Asymmetric Configuration for CCS-MLI

In order to synthesize a multilevel waveform with equal step-size, the CCS-MLI topology needs to use a natural number sequence of sources, i.e., V_{DC}, $2V_{DC}$, $3V_{DC}$, ..., NV_{DC}. Also, placing the sources at the correct positions is equally important. It is determined that the position of sources has to be as shown in Table 6.7 (shown for up to 10 input DC sources).

Accordingly, an algorithm for asymmetric source configuration in the CCS-MLI with N number of input DC sources would be:

1. For an even number of sources:

$$\begin{aligned} V_{DC,j} &= (2j-1)V_{DC}, \text{ for } 1 \leq j \leq (N/2), \\ &= 2(N+1-j)V_{DC}, \text{ for } [(N+2)/2] \leq j \leq N \end{aligned} \tag{6.13}$$

2. For an odd number of sources:

$$\begin{aligned} V_{DC,j} &= (2j-1)V_{DC}, \text{ for } 1 \leq j \leq [(N+1)/2], \\ &= 2(N+1-j)V_{DC}, \text{ for } [(N+3)/2] \leq j \leq N \end{aligned} \tag{6.14}$$

Moreover, the number of levels in the output waveform would be

$$n_{\text{level,asymmetric}} = N^2 + N + 1 \tag{6.15}$$

And the amplitude of the output voltage would be

$$v_{\text{max,asymmetric}} = V_{DC} \sum_{j=1}^{N} j \tag{6.16}$$

To illustrate employment of asymmetry proposed in Eqs. (6.13) and (6.14), four input DC sources of values 6, 12, 18, and 24 V are taken for the CCS-MLI, with the structure as shown in Fig. 6.19A.

The expected number of levels can be calculated using Eq. (6.15) with $N=4$. A 21-level output waveform is expected from the structure with a peak voltage of 60 V as calculated from Eq. (6.16). Also, the waveform should have equal-sized steps of 6 V each. A multicarrier phase opposition disposition (POD) modulation scheme with carrier waveforms of 100-Hz frequency and sinusoidal reference of 50-Hz frequency is used to generate the switching signals. The simulated voltage waveform is shown in Fig. 6.19B. As expected, the waveform consists of 21 levels in equal steps of 6 V with an amplitude of 60 V. The experimentally obtained output voltage waveform is shown in Fig. 6.19C, which exhibits all features as predicted in theory and obtained through simulation.

It can also be noted that for a 21-level output, the CCS-MLI topology would require 22 switches with symmetric source configuration and 10 switches with the asymmetric source algorithm. Thus, asymmetric source

TABLE 6.7 Placement of Sources in CCS-MLI Topology with Asymmetric Configuration

Number of Sources	Source Designation and Their Values									
	$V_{DC,1}$	$V_{DC,2}$	$V_{DC,3}$	$V_{DC,4}$	$V_{DC,5}$	$V_{DC,6}$	$V_{DC,7}$	$V_{DC,8}$	$V_{DC,9}$	$V_{DC,10}$
One	V_{DC}	–	–	–	–	–	–	–	–	–
Two	V_{DC}	$2V_{DC}$	–	–	–	–	–	–	–	–
Three	V_{DC}	$3V_{DC}$	$2V_{DC}$	–	–	–	–	–	–	–
Four	V_{DC}	$3V_{DC}$	$4V_{DC}$	$2V_{DC}$	–	–	–	–	–	–
Five	V_{DC}	$3V_{DC}$	$5V_{DC}$	$4V_{DC}$	$2V_{DC}$	–	–	–	–	–
Six	V_{DC}	$3V_{DC}$	$5V_{DC}$	$6V_{DC}$	$4V_{DC}$	$2V_{DC}$	–	–	–	–
Seven	V_{DC}	$3V_{DC}$	$5V_{DC}$	$7V_{DC}$	$6V_{DC}$	$4V_{DC}$	$2V_{DC}$	–	–	–
Eight	V_{DC}	$3V_{DC}$	$5V_{DC}$	$7V_{DC}$	$8V_{DC}$	$6V_{DC}$	$4V_{DC}$	$2V_{DC}$	–	–
Nine	V_{DC}	$3V_{DC}$	$5V_{DC}$	$7V_{DC}$	$9V_{DC}$	$8V_{DC}$	$6V_{DC}$	$4V_{DC}$	$2V_{DC}$	–
Ten	V_{DC}	$3V_{DC}$	$5V_{DC}$	$7V_{DC}$	$9V_{DC}$	$10V_{DC}$	$8V_{DC}$	$6V_{DC}$	$4V_{DC}$	$2V_{DC}$

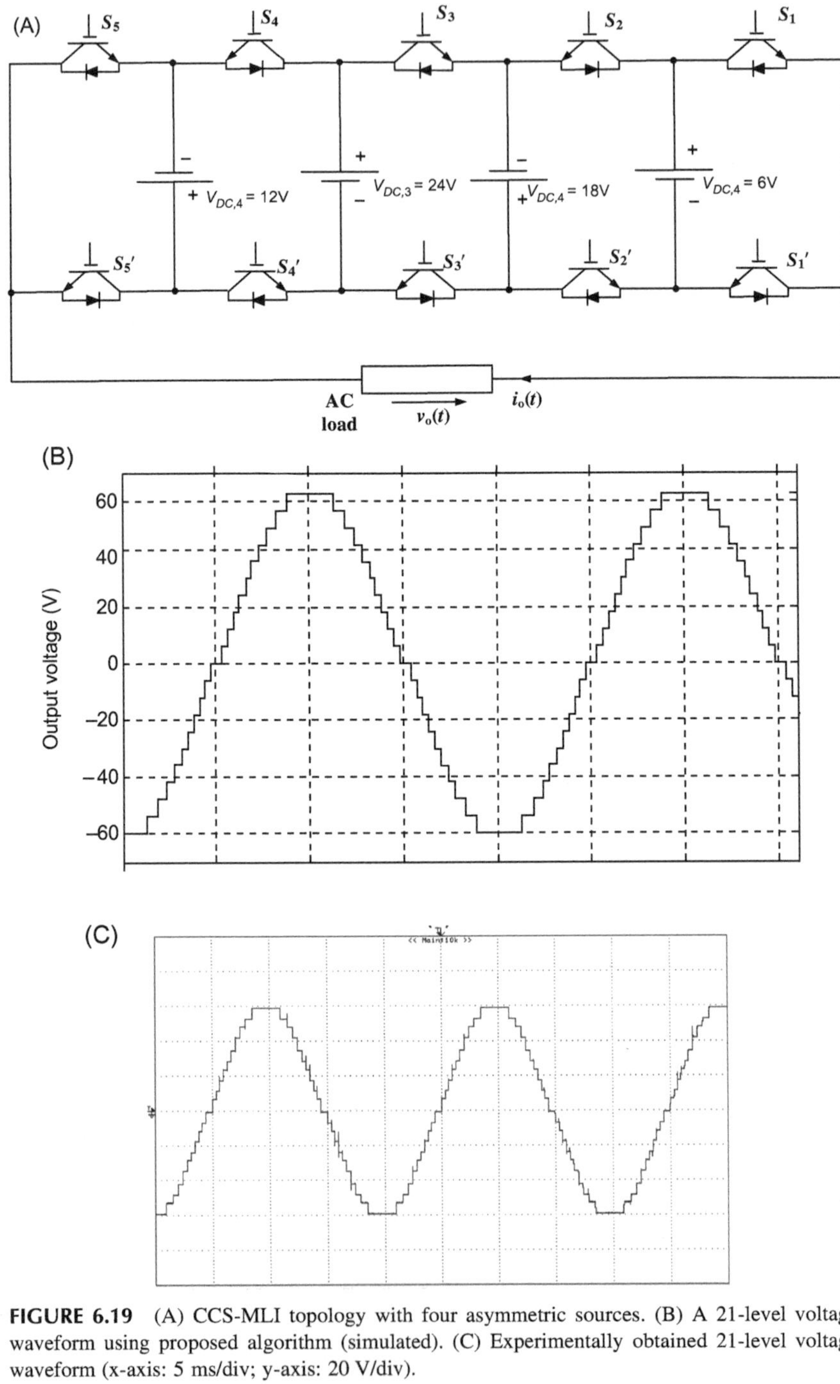

FIGURE 6.19 (A) CCS-MLI topology with four asymmetric sources. (B) A 21-level voltage waveform using proposed algorithm (simulated). (C) Experimentally obtained 21-level voltage waveform (x-axis: 5 ms/div; y-axis: 20 V/div).

configuration would further reduce the device count. However, it comes at the cost of increased power ratings of switches as evident from Eq. (6.8).

6.6 CASCADED MULTILEVEL INVERTER USING SERIES CONNECTION OF CROSS-CONNECTED SOURCES-BASED SUBMULTILEVEL INVERTERS

In the previous sections of this chapter, the so-called cross-connected sources-based multilevel inverter (CCS-MLI) was presented in its generalized form. For a number of input sources greater than two, the following limitations are observed in the generalized structure:

1. Power switches with higher-voltage stresses do not operate at fundamental switching frequency, although their switching frequencies are much lower than the carrier frequency.
2. In the case of symmetrical source configuration, the input sources do not share the load equally.
3. In the case of asymmetrical source configuration, popular configurations like binary and trinary are not applicable. A natural number increment (an *arithmetic progression*) in source voltages is employed, thus depriving the benefits of a *geometric progression* (e.g., binary and trinary).

Overcoming the aforementioned deficiencies would be especially beneficial for medium/high-power and high-voltage applications with high-resolution voltage waveforms. An approach to obtain a high-quality waveform (i.e., a large number of levels), is to employ series-connected submodules/submultilevel inverters to obtain a cascaded multilevel inverter [6–21], as illustrated in Fig. 6.20 for a single-phase structure.

In general, cascaded structures offer the following advantages [6,20,21]:

1. Modularity and scalability;
2. Flexibility in converter design, adapting to different ranges of voltages and powers; the possibility of using various multilevel submodule

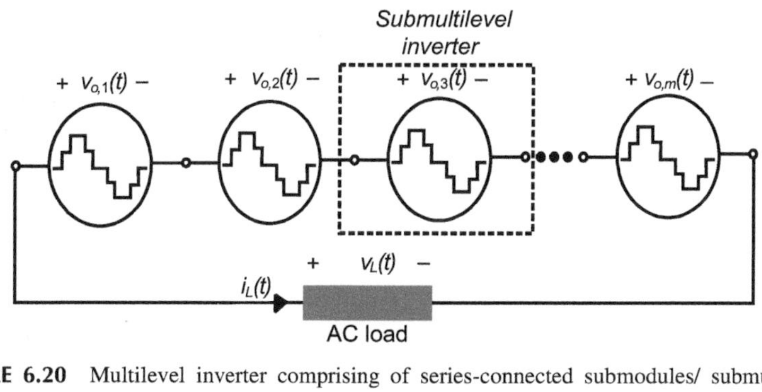

FIGURE 6.20 Multilevel inverter comprising of series-connected submodules/ submultilevel inverters.

concepts to incorporate well-proven multilevel submodule designs to novel cascaded designs;
3. Possibility of an effective fault-tolerant operation by simply bypassing the faulted "cell(s)"; and
4. Reducing the dimensions of the whole converter at the expense of complicating the technology of the submodule.

In the light of discussions on the limitations of CCS-MLI with more than two input sources and the advantages of cascaded multilevel structures using submultilevel inverters, discussions are carried out in subsequent sections of this chapter to use CCS-MLI with two (symmetric or asymmetric) input sources as a submodule/submultilevel inverter to make a cascaded multilevel inverter. Each submodule with two input sources is referred to as a "cell." Symmetric and asymmetric source configurations are separately investigated.

6.7 CASCADED MULTILEVEL INVERTER WITH SYMMETRIC SOURCES

The symmetric cascaded multilevel inverter is shown in Fig. 6.21. The basic building block, or "cell," of this structure is a five-level inverter using two

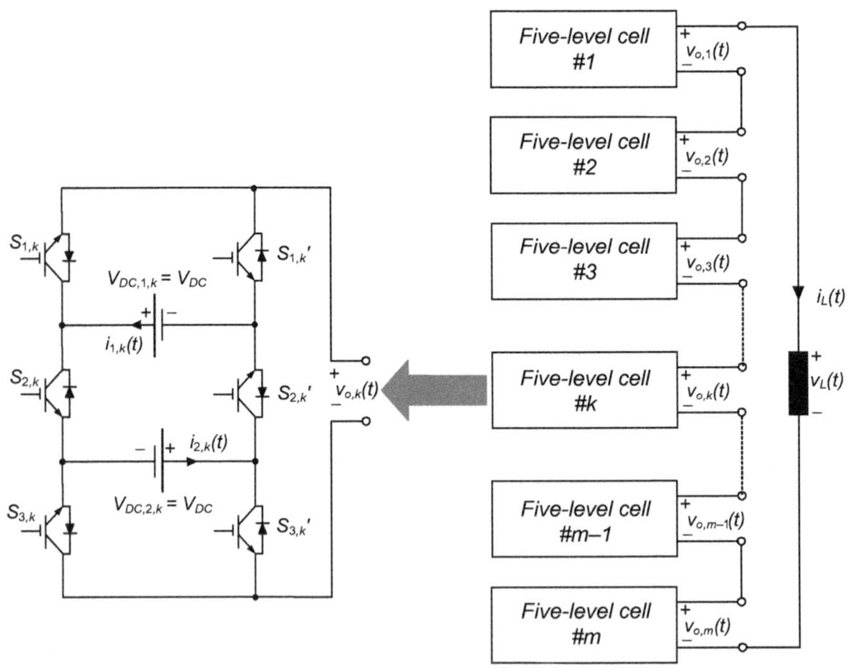

FIGURE 6.21 Cascaded multilevel inverter comprising of series-connected five-level cells with symmetric input sources.

cross-connected equal sources. "m" number of five-level cells are connected in series to obtain the cascaded inverter. The kth cell $\{k=1 \text{ to } m\}$ consists of two input DC sources and six power switches.

As discussed in previous sections of this chapter, each cell exhibits the following features:

1. For sources $V_{DC,1,k}=V_{DC,2,k}=V_{DC}$ $\{k=1 \text{ to } m\}$, four switches tolerate a voltage stress of V_{DC} each and two switches tolerate a voltage stress of $2V_{DC}$ each. With proper control strategy, switches with a higher voltage stress of $2V_{DC}$ always operate at a fundamental switching frequency *irrespective of the carrier frequency*.
2. With proper control strategy, power balancing takes place between the two equal sources within two cycles of output voltage.

For a structure with m number of five-level cells, the number of levels in the output voltage will be given by,

$$n_{\text{levels,symmetric}} = 4m + 1 \tag{6.17}$$

Also, the structure would require $2m$ number of sources with $6m$ number of switches, of which $2m$ switches block a voltage of $2V_{DC}$ and $4m$ switches block a voltage of V_{DC}. The load voltage is given by,

$$v_L(t) = \sum_{k=1}^{m} v_{o,k}(t) = V_{DC}\sum_{k=1}^{m}[T_{1,k}(t) - 2T_{2,k}(t) + T_{3,k}(t)] \tag{6.18}$$

where, T corresponds to the switching function of a given switch S such that,

$$T(t) = \begin{cases} 1, & \text{when switch } S \text{ is ON,} \\ 0, & \text{when switch } S \text{ is OFF} \end{cases} \tag{6.19}$$

6.7.1 Control Scheme

The cascaded inverter can be controlled with the scheme as explained in Chapter 4, Universal Control Scheme With Voltage-Level-Based Methods. However, in order to preserve the modular nature of the cascaded structure, it is desirable that each five-level cell is identically modulated. Moreover, it is also desired that both the high-voltage switches in a cell operate at fundamental frequency and input sources share load equally. Thus, the modulation strategy described in Section 6.4.2 for a five-level inverter is replicated for all cells with appropriately phase-shifted carrier signals being used in each cell. That is, for a given k^{th} cell $\{k=1 \text{ to } m\}$, there are four carrier signals, as shown in Fig. 6.22, each carrier with a phase shift of $\frac{360 \text{ degrees}}{m}$ with respect to the corresponding carrier of each cell, thus the carrier configuration can be selected from many options as described in Chapter 4, Universal Control Scheme With Voltage-Level-Based Methods.

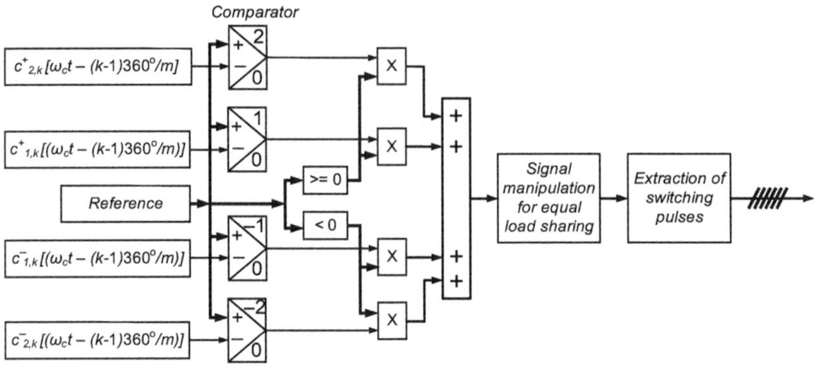

FIGURE 6.22 Control scheme for *k*th cell for cascaded MLI shown in Fig. 6.21.

For example, if there are two cascaded five-level cells, then each carrier of the second cell is 180 degrees phase-shifted with respect to the corresponding carriers used in the first cell. Subsequently, each cell is modulated with an appropriately manipulated aggregated signal as explained in Section 6.4.2. In this manner, while all the cells operate with desired characteristics (i.e., fundamental switching frequency of high-voltage switches and equal load sharing amongst the input sources), phase-shifting of carriers ensures that multilevel waveform is synthesized at the output terminals with $4\,m+1$ number of levels.

6.7.2 Simulation Study

In order to verify the cascaded topology with symmetric sources and the modulation strategy, a nine-level inverter with two cells (i.e., $m=2$) and four sources $V_{DC,1,1}=V_{DC,1,2}=V_{DC,2,1}=V_{DC,2,2}=V_{DC}=40$ V is simulated using MATLAB/Simulink. An inductive load with $R=35\ \Omega$ and $L=3.25$ mH is considered. The power circuit for the structure is shown in Fig. 6.23.

Phase-disposed triangular signals with a frequency of 1 kHz are employed as carriers and sinusoidal reference with a frequency of 50 Hz is used with an amplitude modulation index of 0.85. Switching signals obtained for various switches are shown in Fig. 6.24. The output voltages of the two cells ($v_{o,1}(t)$ and $v_{o,2}(t)$), the load voltage ($v_L(t)$) and the load current ($i_L(t)$) are shown in Fig. 6.25.

It can be observed from Fig. 6.24 that the switches with higher voltage-blocking requirement (switches $S_{2,1}$, $S_{2,1}'$, $S_{2,2}$, and $S_{2,2}'$) operate at a fundamental switching frequency of 50 Hz. Fig. 6.25 indicates that the two cells synthesize five-level voltages, while at the load terminals a nine-level voltage is obtained as expected (as can be verified with $m=2$ in Eq. (6.17)). The source current waveforms are shown in Fig. 6.26. Source current waveforms indicate that the average values of the currents are equal and, since the source voltages are equal, they contribute equal power to the load. It is important to

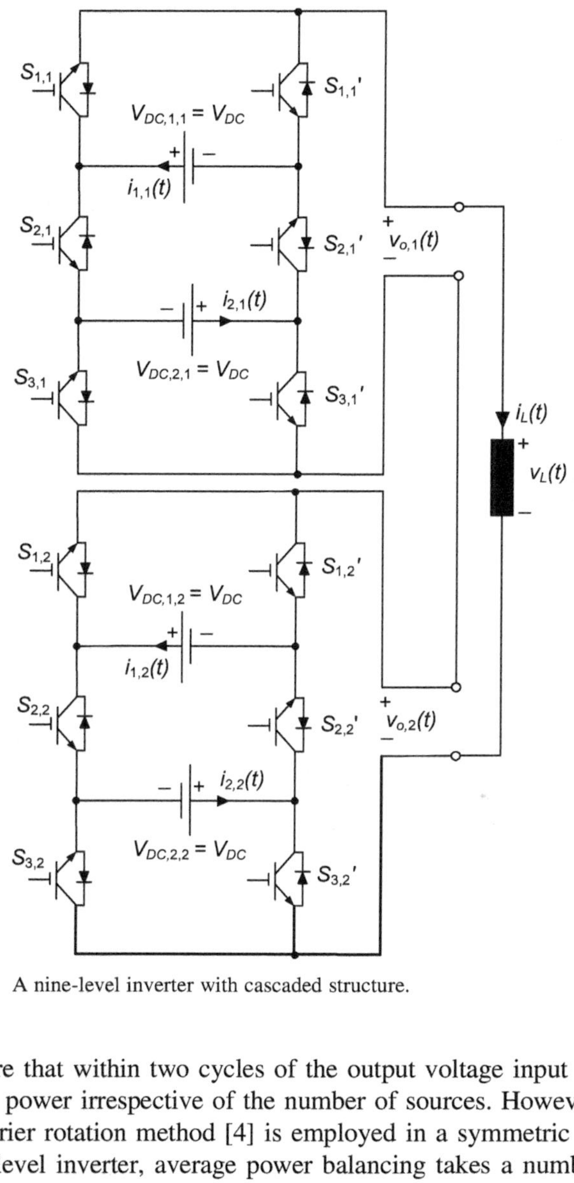

FIGURE 6.23 A nine-level inverter with cascaded structure.

point out here that within two cycles of the output voltage input sources contribute equal power irrespective of the number of sources. However, when the so-called carrier rotation method [4] is employed in a symmetric cascaded H-bridge multilevel inverter, average power balancing takes a number of cycles equal to the number of input sources. Thus, with the discussed topology and modulation scheme, an important advantage is derived.

6.7.3 Experimental Verification

A low-power low-voltage experimental prototype of a nine-level inverter based on the proposed topology has been developed based on the circuit

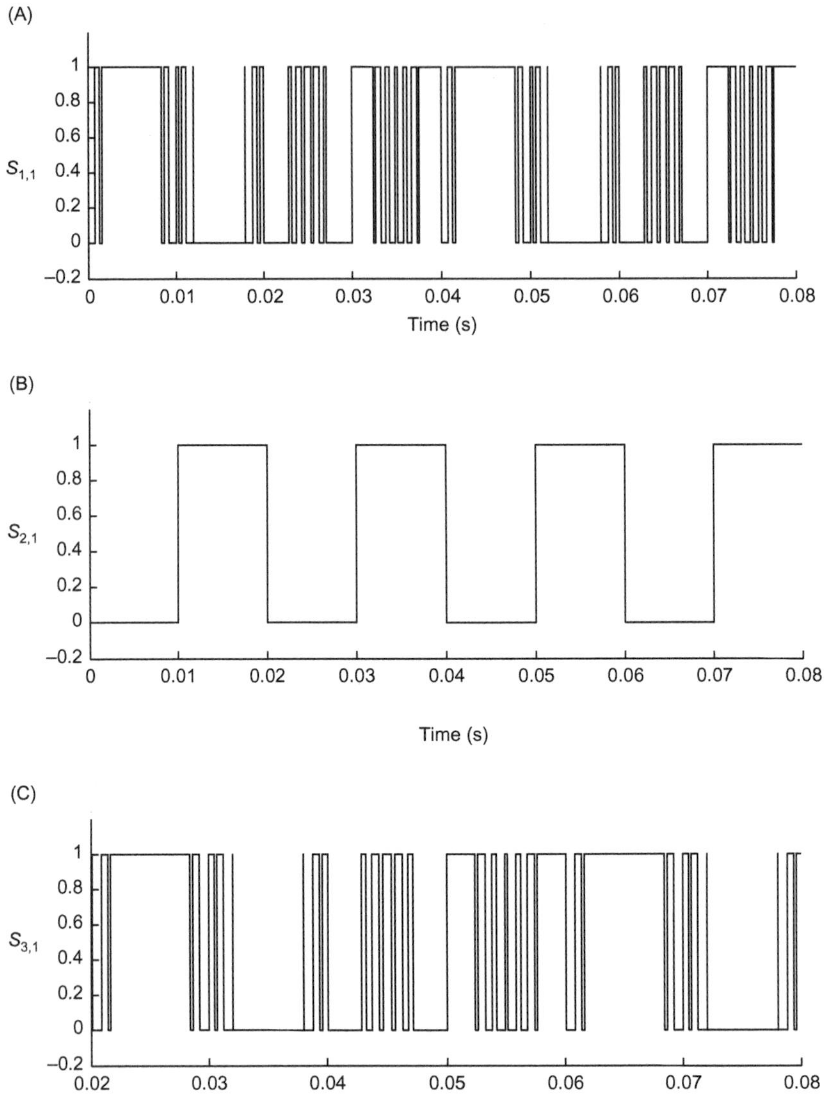

FIGURE 6.24 Switching signals for the nine-level cascaded inverter.

shown in Fig. 6.23. Four regulated power supplies with equal voltages, $V_{DC,1,1} = V_{DC,1,2} = V_{DC,2,1} = V_{DC,2,2} = V_{DC} = 40$ V, are used as input DC levels. A power module with IGBTs STGW30N120KD and appropriate driver circuits are used. Reference and carrier waveforms as described for the simulation study are used and gating signals are obtained using real-time interfacing of Simulink model with dSPACE DS1103 real-time controller. Gate signals for the power switches as obtained from dSPACEDS1103 are

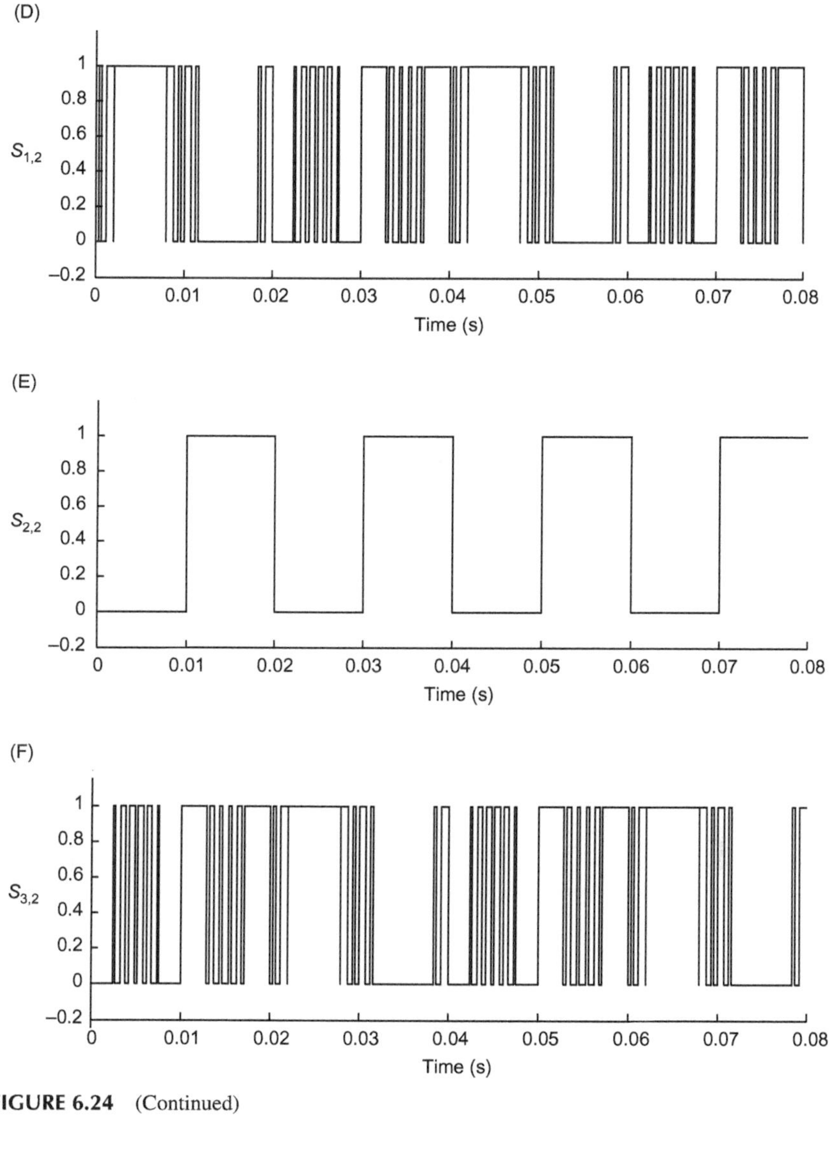

FIGURE 6.24 (Continued)

shown in Fig. 6.27. It can be seen that switches with higher voltage stresses are commutated at a fundamental frequency of 50 Hz.

Voltage waveforms of the two five-level cascaded cells and load are shown in Fig. 6.28. A nine-level load voltage is obtained in equal steps of 40 V.The inverter is loaded with an R–L load of values $R = 35\ \Omega$ and $L = 3.25$ mH. The load current waveform indicates the inductive nature of the load. Also, the source current waveforms for the inverter are shown in Fig. 6.29. It can be observed that the waveforms would have equal average

(A)

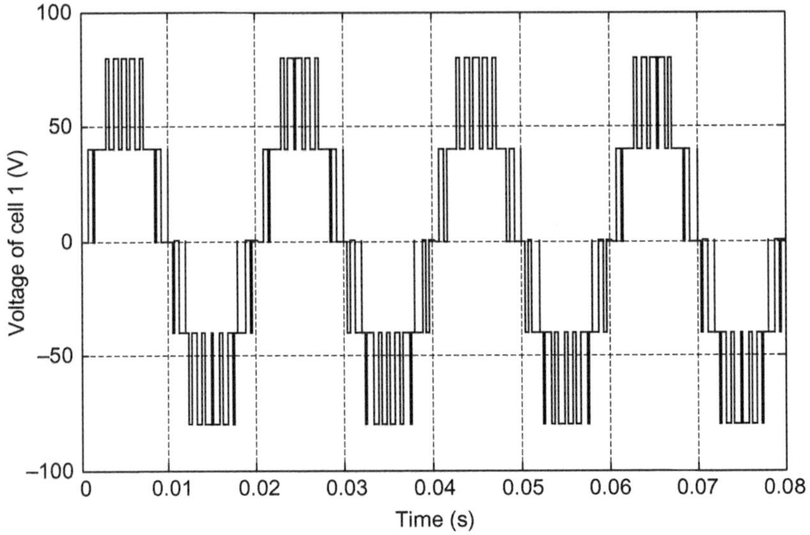

(B)

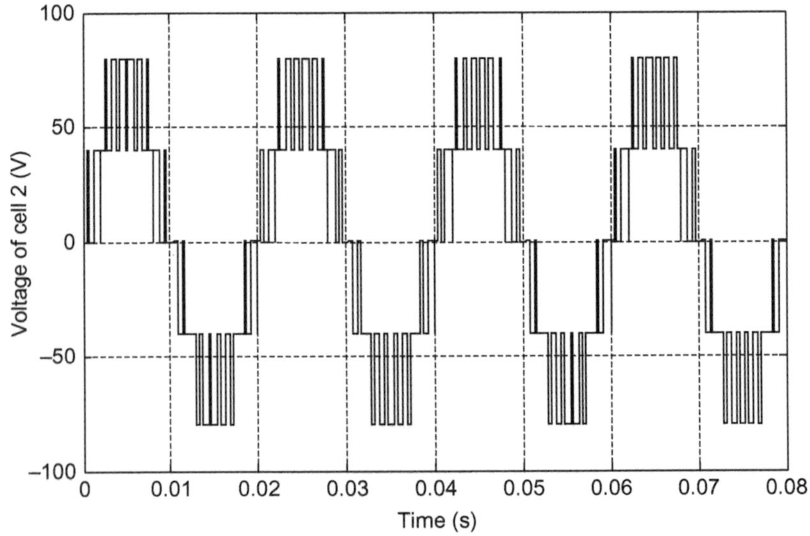

FIGURE 6.25 Simulated waveforms for the nine-level cascaded inverter; (A) output voltage of cell 1 ($v_{0,1}(t)$); (B) output voltage of cell 2 ($v_{0,2}(t)$); (C) load voltage ($v_L(t)$); and (D) load current ($i_L(t)$).

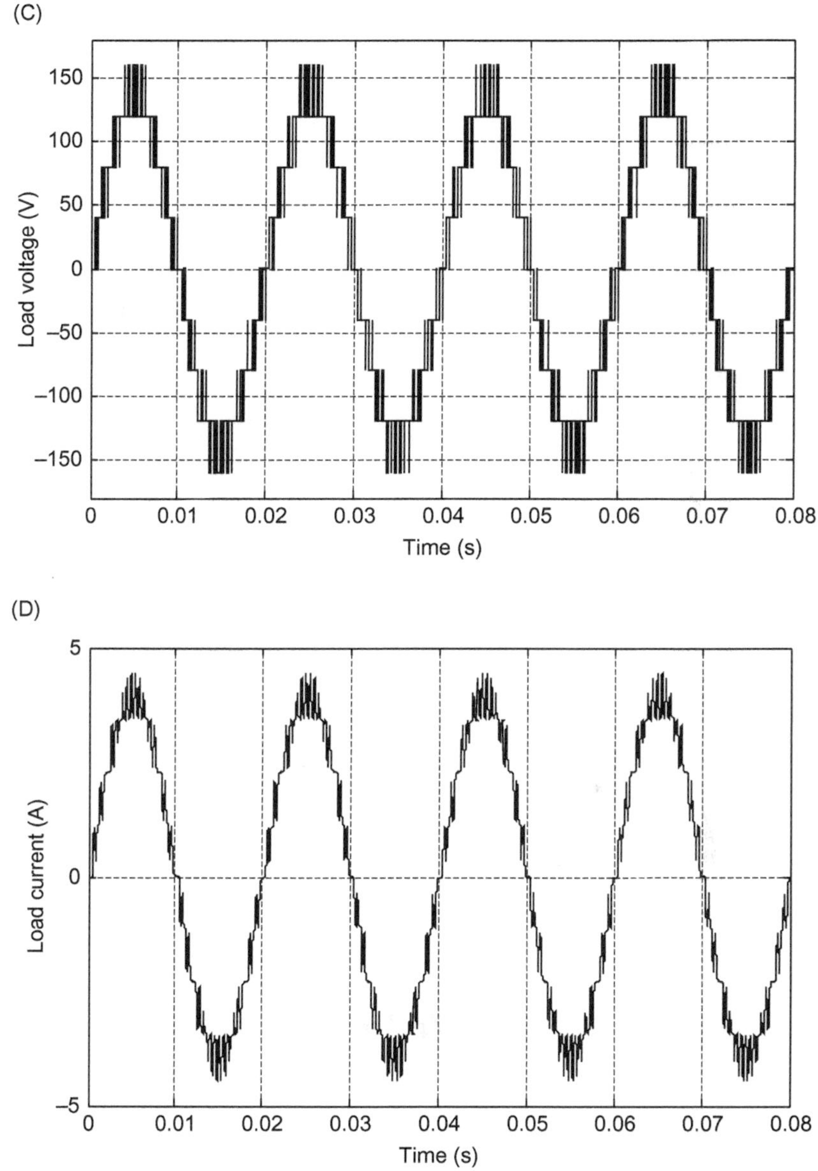

FIGURE 6.25 (Continued)

values over every two cycles of the fundamental period. In addition, power measurements are carried out for different output loads for the four input DC sources and summarized in Table 6.8. It can be observed that all the sources are equally loaded with 72 W each.

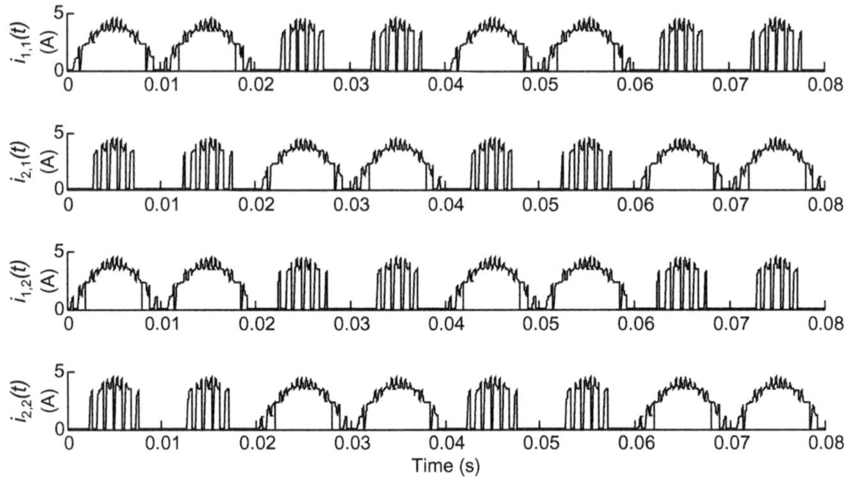

FIGURE 6.26 Source current waveforms for the nine-level cascaded inverter; $i_{1,1}(t)$, $i_{2,1}(t)$, $i_{1,2}(t)$, and $i_{2,2}(t)$.

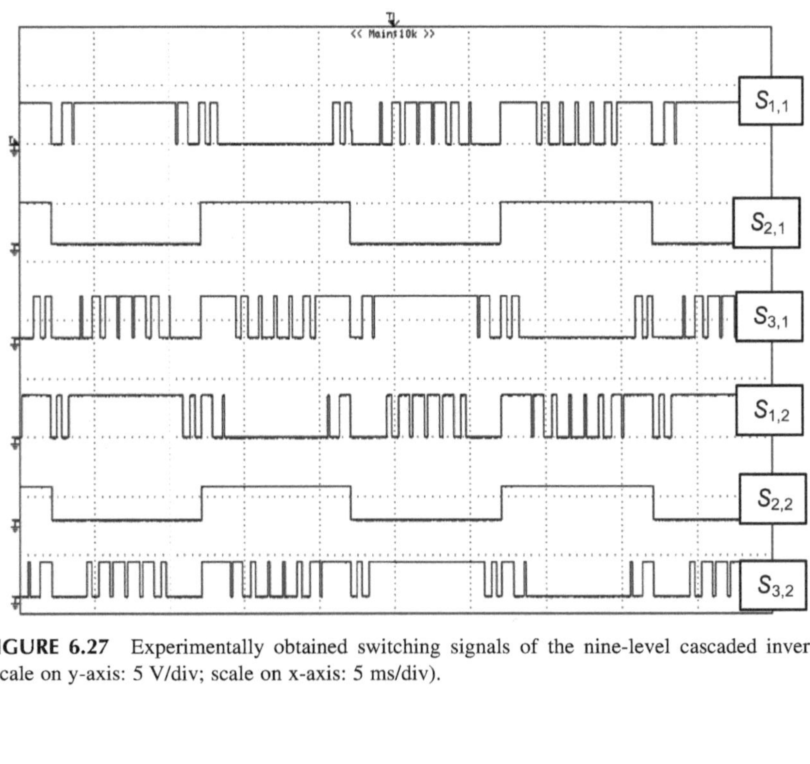

FIGURE 6.27 Experimentally obtained switching signals of the nine-level cascaded inverter (scale on y-axis: 5 V/div; scale on x-axis: 5 ms/div).

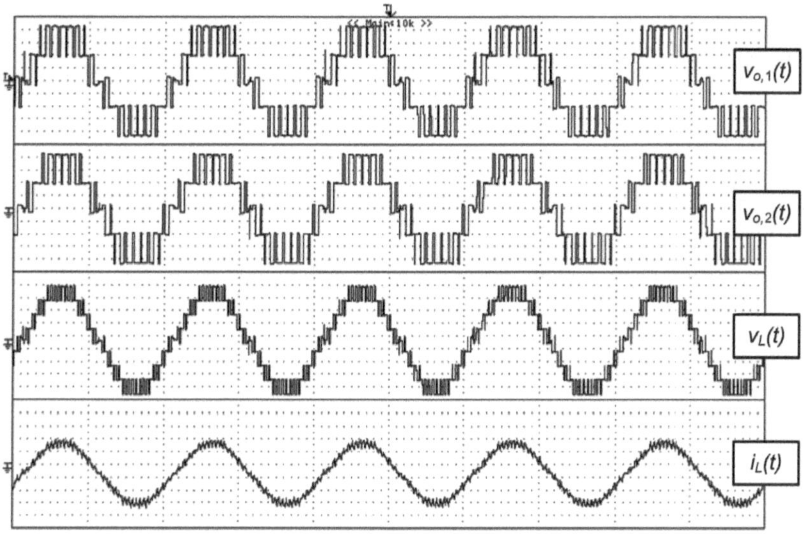

FIGURE 6.28 Experimentally obtained cell voltage, output voltage, and load current waveforms of the cascaded nine-level inverter ($V_{o,1}(t)$: 20 V/div; $V_{o,2}(t)$: 20 V/div; $V_L(t)$: 40 V/div; $i_L(t)$: 2 A/div; scale on x-axis: 5 ms/div).

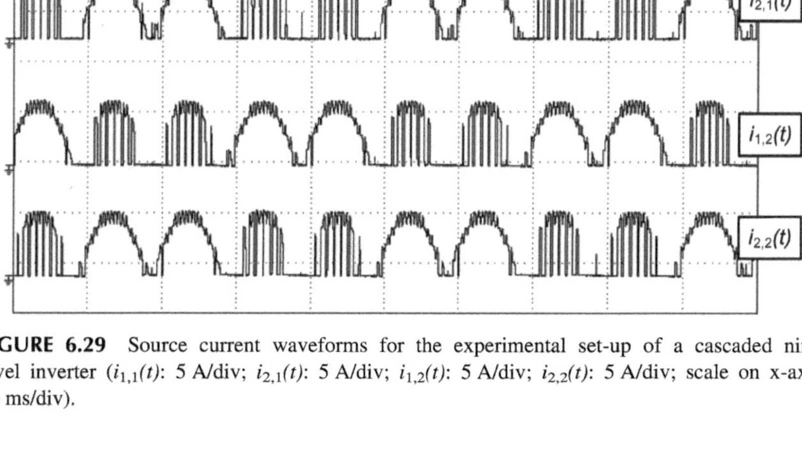

FIGURE 6.29 Source current waveforms for the experimental set-up of a cascaded nine-level inverter ($i_{1,1}(t)$: 5 A/div; $i_{2,1}(t)$: 5 A/div; $i_{1,2}(t)$: 5 A/div; $i_{2,2}(t)$: 5 A/div; scale on x-axis: 10 ms/div).

TABLE 6.8 Measurement of Power at Input Sources and Load for the Experimental Set-up of a Cascaded Nine-Level Inverter

Output Power (W)	Input Power			
	Power Delivered by Source $V_{DC,1}$ (W)	Power Delivered by Source $V_{DC,1}$ (W)	Power Delivered by Source $V_{DC,1}$ (W)	Power Delivered by Source $V_{DC,1}$ (W)
250.0	72.0	72.0	72.0	72.0

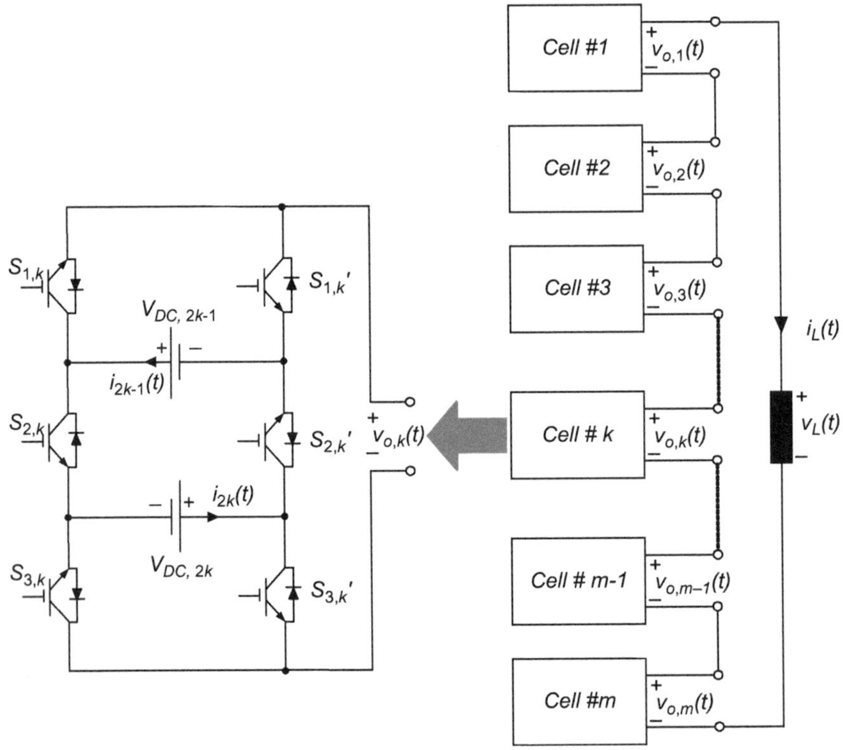

FIGURE 6.30 Cascaded multilevel inverter comprising of series-connected cells with asymmetric input sources.

6.8 CASCADED MULTILEVEL INVERTER WITH ASYMMETRIC SOURCES

The asymmetric cascaded multilevel inverter based on CCS-MLI is shown in Fig. 6.30.

The basic building block, or "cell," of this structure is a multilevel inverter with two cross-connected sources, $V_{DC,2k-1}$ and $V_{DC,2k}$, $\{k=1$ to $m\}$. "m" number of cells are connected in series to obtain the proposed cascaded inverter. The kth cell $\{k=1$ to $m\}$ consists of two input DC sources and six power switches. For employing asymmetric source configuration, the fundamental switching frequency operation of the switch pairs $(S_{2,k},S_{2,k}')$ is to be preserved. The output voltage of the kth cell is given by

$$v_{o,k}(t)=V_{DC,2k-1}\{T_{1,k}(t)-T_{2,k}(t)\}-V_{DC,2k}\{T_{2,k}(t)-T_{3,k}(t)\} \tag{6.20}$$

Also, the output voltage at the load terminals is given by

$$v_L(t)=\sum_{k=1}^{m}v_o(t)=\sum_{k=1}^{m}\{V_{DC,2k-1}\{T_{1,k}(t)-T_{2,k}(t)\}-V_{DC,2k}\{T_{2,k}(t)-T_{3,k}(t)\}\} \tag{6.21}$$

Eqs. (6.20) and (6.21) are the guiding expressions to obtain appropriate voltage source ratios so as to synthesize the output voltage with equal-sized steps. Thus, with the constraints of fundamental switching of switch pairs $(S_{2,k}, S_{2,k}')$ $\{k=1$ to $m\}$, equal-sized steps in the output waveform and input source voltage ratios with geometric progression, the best-suited source asymmetry amongst many possibilities would be "binary" [12]. That is to say, source values will follow the algorithm,

$$V_{DC,2k-1}=2^{2k-2}V_{DC} \tag{6.22}$$

Accordingly, the number of levels in the output voltage will be

$$n_{\text{levels,asymmetric}}=2^{2m+1}-1 \tag{6.23}$$

As an example, a cascaded inverter with two cells is shown in Fig. 6.31.

6.8.1 Control Scheme

The asymmetric source-cascaded inverter can be controlled with the scheme as explained in Chapter 4, Universal Control Scheme With Voltage-Level-Based Methods. However, it is desired that both the high-voltage switches in a cell operate at fundamental frequency. Thus, the modulation strategy described in Section 6.4.2, which targets optimal switching operation of a five-level inverter based on CCS-MLI, is further extended for the asymmetric cascaded structure. For the power circuit with $m=2$ as shown in Fig. 6.31, the control scheme is illustrated in Fig. 6.32.

Fundamental switching of higher-voltage switches of a cell entails both the cells simultaneously synthesizing their respective positive voltage levels and/or zero levels for the positive half cycle of the output voltage. Similarly, both cells should simultaneously synthesize their respective negative voltage levels and/or zero levels. The look-up table for the structure is shown as Table 6.9.

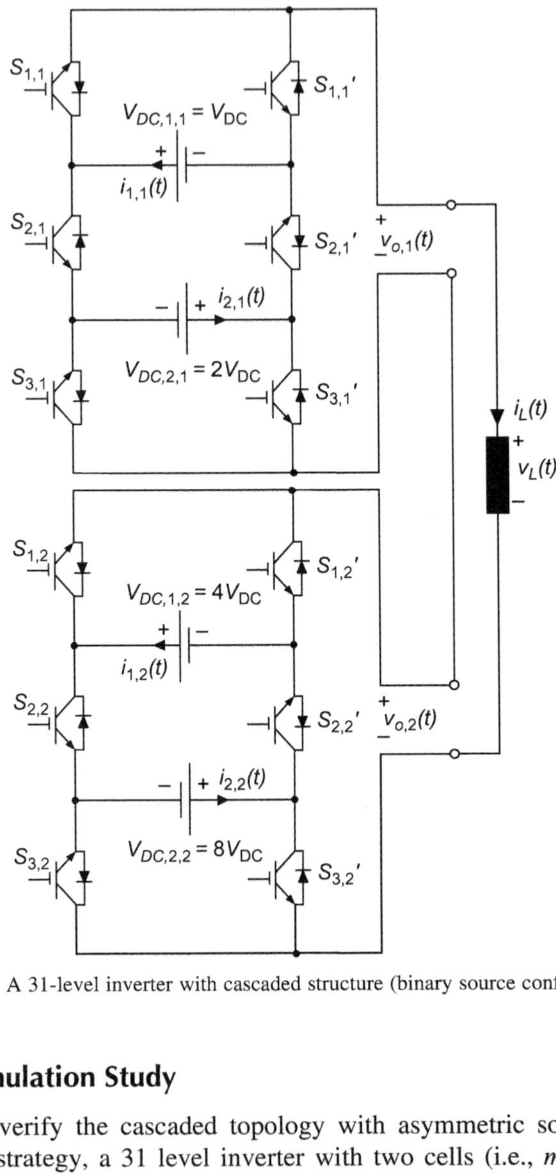

FIGURE 6.31 A 31-level inverter with cascaded structure (binary source configuration).

6.8.2 Simulation Study

In order to verify the cascaded topology with asymmetric sources and the modulation strategy, a 31 level inverter with two cells (i.e., $m=2$) and four sources $V_{DC,1,1} = 15V$, $V_{DC,1,2} = 30V$, $V_{DC,2,1} = 60V$ and $V_{DC,2,2} = 120V$ is simulated using MATLAB/Simulink. An inductive load is considered ($R = 30\Omega$ and $L = 3.25$ mH). Power circuit for the structure is shown in Fig. 6.31. Control scheme and look-up table as described in the previous section are used. Phase-disposed triangular signals with frequency 1 kHz are

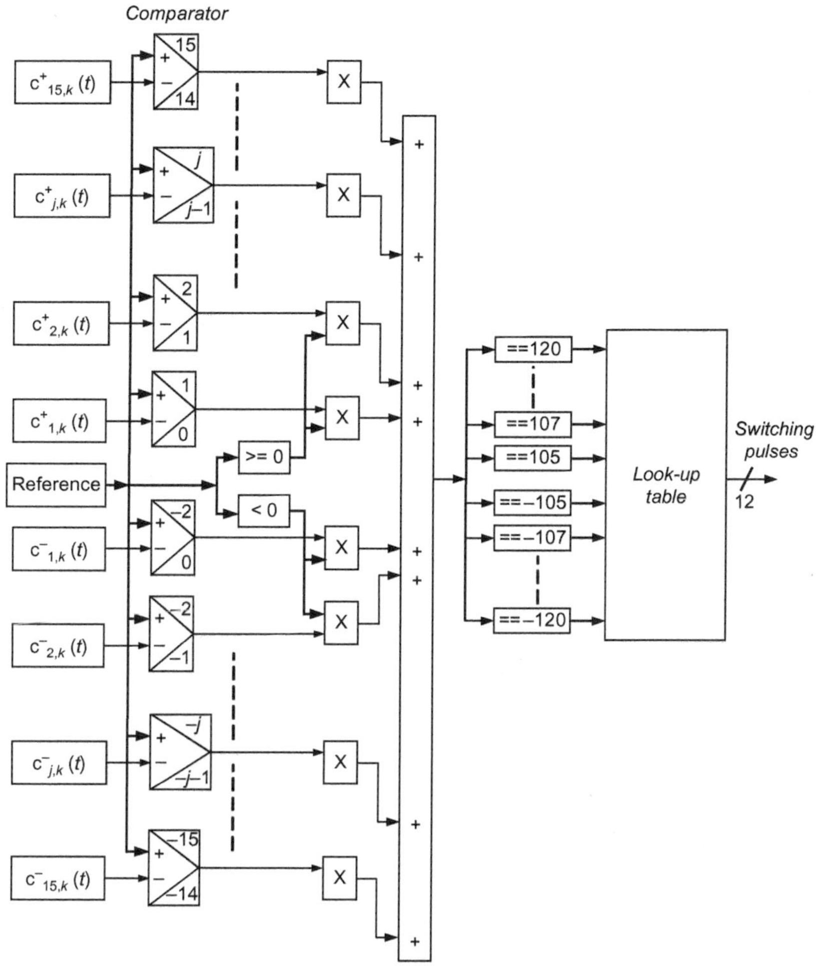

FIGURE 6.32 Control scheme for asymmetric cascaded structure with two cells and binary sources.

employed as carriers and sinusoidal reference with frequency 50 Hz is used with an amplitude modulation index of 0.95. Switching signals as obtained for various switches are shown in Fig. 6.33.

It can be observed from Fig. 6.33 that the switches with a higher voltage-blocking requirement (switches $S_{2,1}$, $S'_{2,1}$, $S_{2,2}$, and $S'_{2,2}$) operate at a fundamental switching frequency of 50 Hz. The output voltages of the two cells and the load voltage are shown in Fig. 6.34. It can be seen that at the load terminals a 31-level voltage is obtained as expected (as can be verified with $m = 2$ in Eq. (6.23)).

TABLE 6.9 Look-Up Table for 31-Level Asymmetric Cascaded Inverter

S. No.	Level of Aggregated Signal "$a(t)$"	$v_{o,1}(t)$	$v_{o,2}(t)$	$v_L(t)$	Switches in ON State (Other Switches Remain OFF)	
1	105	0	0	0	S'_1, S'_2, S'_3	S'_4, S'_5, S'_6
2	106	V_{DC}	0	V_{DC}	S_1, S'_2, S'_3	S'_4, S'_5, S'_6
3	107	$2V_{DC}$	0	$2V_{DC}$	S'_1, S'_2, S_3	S'_4, S'_5, S'_6
4	108	$3V_{DC}$	0	$3V_{DC}$	S_1, S'_2, S_3	S'_4, S'_5, S'_6
5	109	0	$4V_{DC}$	$4V_{DC}$	S'_1, S'_2, S'_3	S_4, S'_5, S'_6
6	110	V_{DC}	$4V_{DC}$	$5V_{DC}$	S_1, S'_2, S'_3	S_4, S'_5, S'_6
7	111	$2V_{DC}$	$4V_{DC}$	$6V_{DC}$	S'_1, S'_2, S_3	S_4, S'_5, S'_6
8	112	$3V_{DC}$	$4V_{DC}$	$7V_{DC}$	S_1, S'_2, S_3	S_4, S'_5, S'_6
9	113	0	$8V_{DC}$	$8V_{DC}$	S'_1, S'_2, S'_3	S'_4, S'_5, S_6
10	114	V_{DC}	$8V_{DC}$	$9V_{DC}$	S_1, S'_2, S'_3	S'_4, S'_5, S_6
11	115	$2V_{DC}$	$8V_{DC}$	$10V_{DC}$	S'_1, S'_2, S_3	S'_4, S'_5, S_6
12	116	$3V_{DC}$	$8V_{DC}$	$11V_{DC}$	S_1, S'_2, S'_3	S'_4, S'_5, S_6
13	117	0	$12V_{DC}$	$12V_{DC}$	S'_1, S'_2, S'_3	S_4, S'_5, S_6
14	118	V_{DC}	$12V_{DC}$	$13V_{DC}$	S_1, S'_2, S'_3	S_4, S'_5, S_6
15	119	$2V_{DC}$	$12V_{DC}$	$14V_{DC}$	S'_1, S'_2, S_3	S_4, S'_5, S_6
16	120	$3V_{DC}$	$12V_{DC}$	$15\ V_{DC}$	S_1, S'_2, S_3	S_4, S'_5, S_6
17	− 105	0	0	0	S_1, S_2, S_3	S_4, S_5, S_6
18	− 106	$-V_{DC}$	0	$-V_{DC}$	S'_1, S_2, S_3	S_4, S_5, S_6
19	− 107	$-2V_{DC}$	0	$-2V_{DC}$	S_1, S_2, S'_3	S_4, S_5, S_6
20	− 108	$-3V_{DC}$	0	$-3V_{DC}$	S'_1, S_2, S'_3	S_4, S_5, S_6
21	− 109	0	$-4V_{DC}$	$-4V_{DC}$	S_1, S_2, S_3	S'_4, S_5, S_6
22	− 110	$-V_{DC}$	$-4V_{DC}$	$-5V_{DC}$	S'_1, S_2, S_3	S'_4, S_5, S_6
23	− 111	$-2V_{DC}$	$-4V_{DC}$	$-6V_{DC}$	S_1, S_2, S'_3	S'_4, S_5, S_6
24	− 112	$-3V_{DC}$	$-4V_{DC}$	$-7V_{DC}$	S'_1, S_2, S'_3	S'_4, S_5, S_6
25	− 113	0	$-8V_{DC}$	$-8V_{DC}$	S_1, S_2, S_3	S_4, S_5, S'_6
26	− 114	$-V_{DC}$	$-8V_{DC}$	$-9V_{DC}$	S'_1, S_2, S_3	S_4, S_5, S'_6
27	− 115	$-2V_{DC}$	$-8V_{DC}$	$-10V_{DC}$	S_1, S_2, S'_3	S_4, S_5, S'_6
28	− 116	$-3V_{DC}$	$-8V_{DC}$	$-11V_{DC}$	S'_1, S_2, S'_3	S_4, S_5, S'_6
29	− 117	0	$-12V_{DC}$	$-12V_{DC}$	S_1, S_2, S_3	S'_4, S_5, S'_6
30	− 118	$-V_{DC}$	$-12V_{DC}$	$-13V_{DC}$	S'_1, S_2, S_3	S'_4, S_5, S'_6
31	− 119	$-2V_{DC}$	$-12V_{DC}$	$-14V_{DC}$	S_1, S_2, S'_3	S'_4, S_5, S'_6
32	− 120	$-3V_{DC}$	$-12V_{DC}$	$-15V_{DC}$	S'_1, S_2, S'_3	S'_4, S_5, S'_6

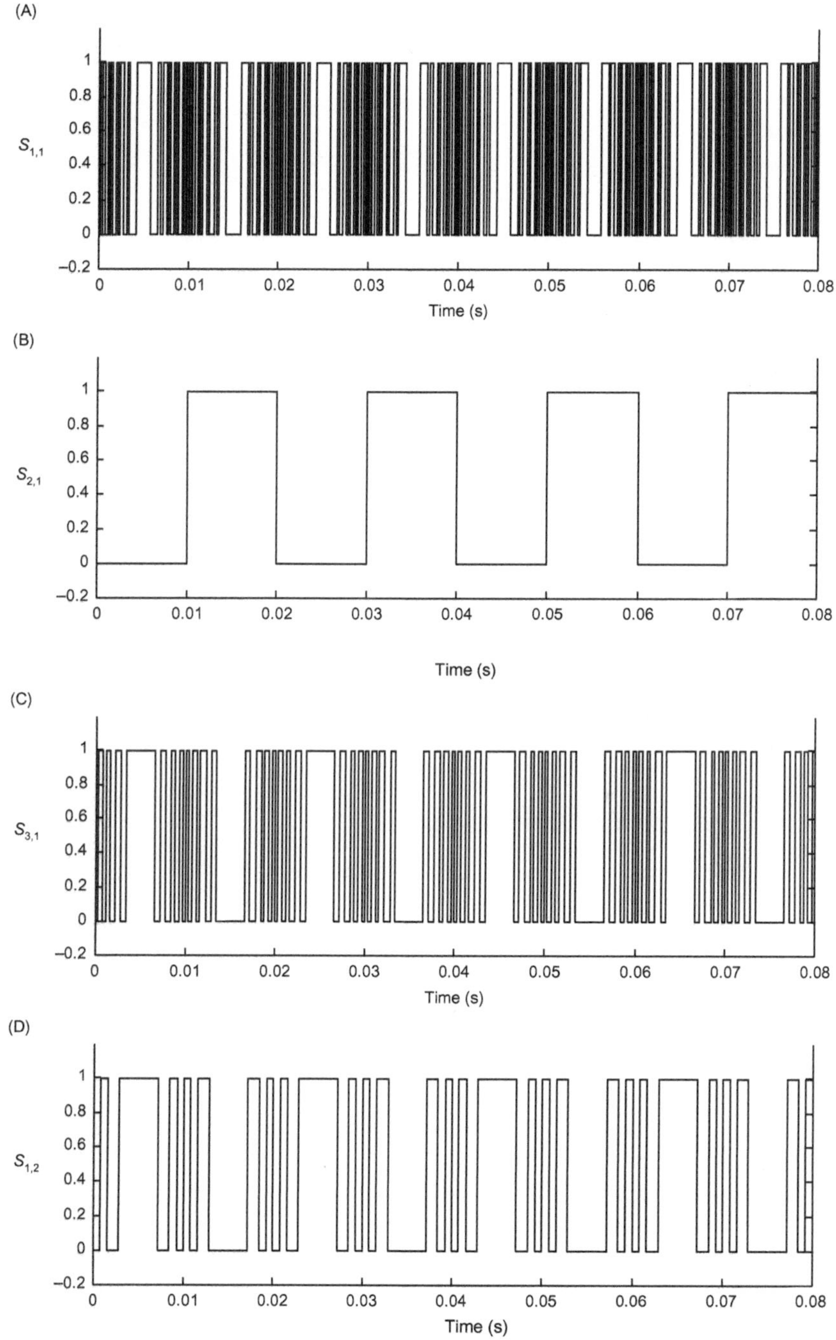

FIGURE 6.33 Switching signals for the 31-level asymmetric cascaded inverter with two cells.

(E)

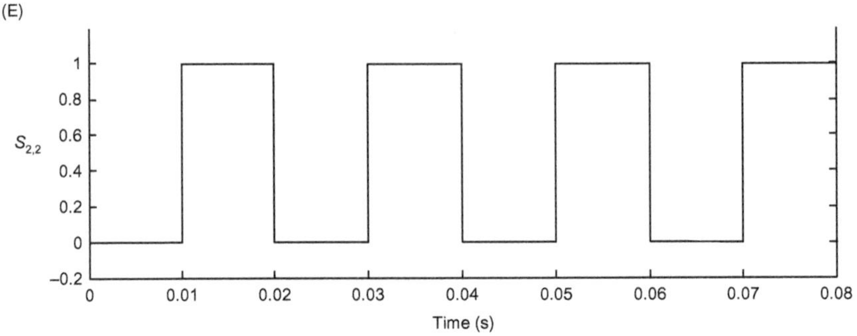

(F)

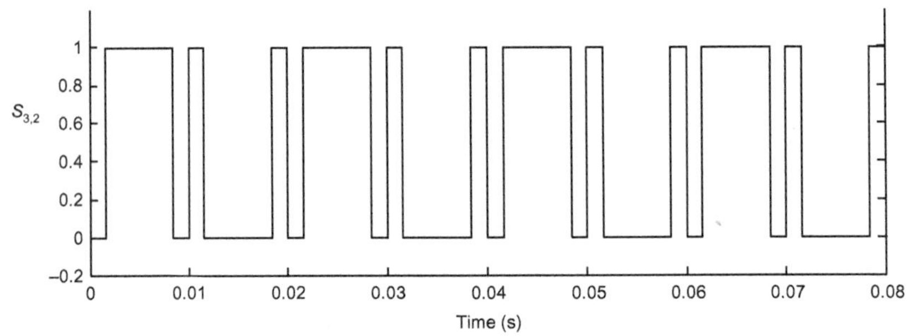

FIGURE 6.33 (Continued)

6.8.3 Experimental Verification

An experimental prototype of a 31-level inverter based on the proposed topology has been developed based on the circuit shown in Fig. 6.31. Four DC supply sources with respective voltages of $V_{DC,1,1} = 15$ V, $V_{DC,1,2} = 30$ V, $V_{DC,2,1} = 60$ V, and $V_{DC,2,2} = 120$ V are used as input DC levels. A power module with IGBTs (STGW30N120KD) and appropriate driver circuits are used. Reference and carrier waveforms as described for the simulation study are used and gating signals are generated through a dSPACE DS1103 real-time controller. Gate signals so obtained are shown in Fig. 6.35. It can be seen that switches with higher voltage stresses are operated at a fundamental frequency of 50 Hz.

Voltage waveforms of the two cascaded cells, load voltage, and load current waveforms are shown in Fig. 6.36. A 31-level load voltage is obtained in equal steps of 15 V. The load current waveform indicates the inductive nature of the load and the waveforms are in close agreement with the simulation results.

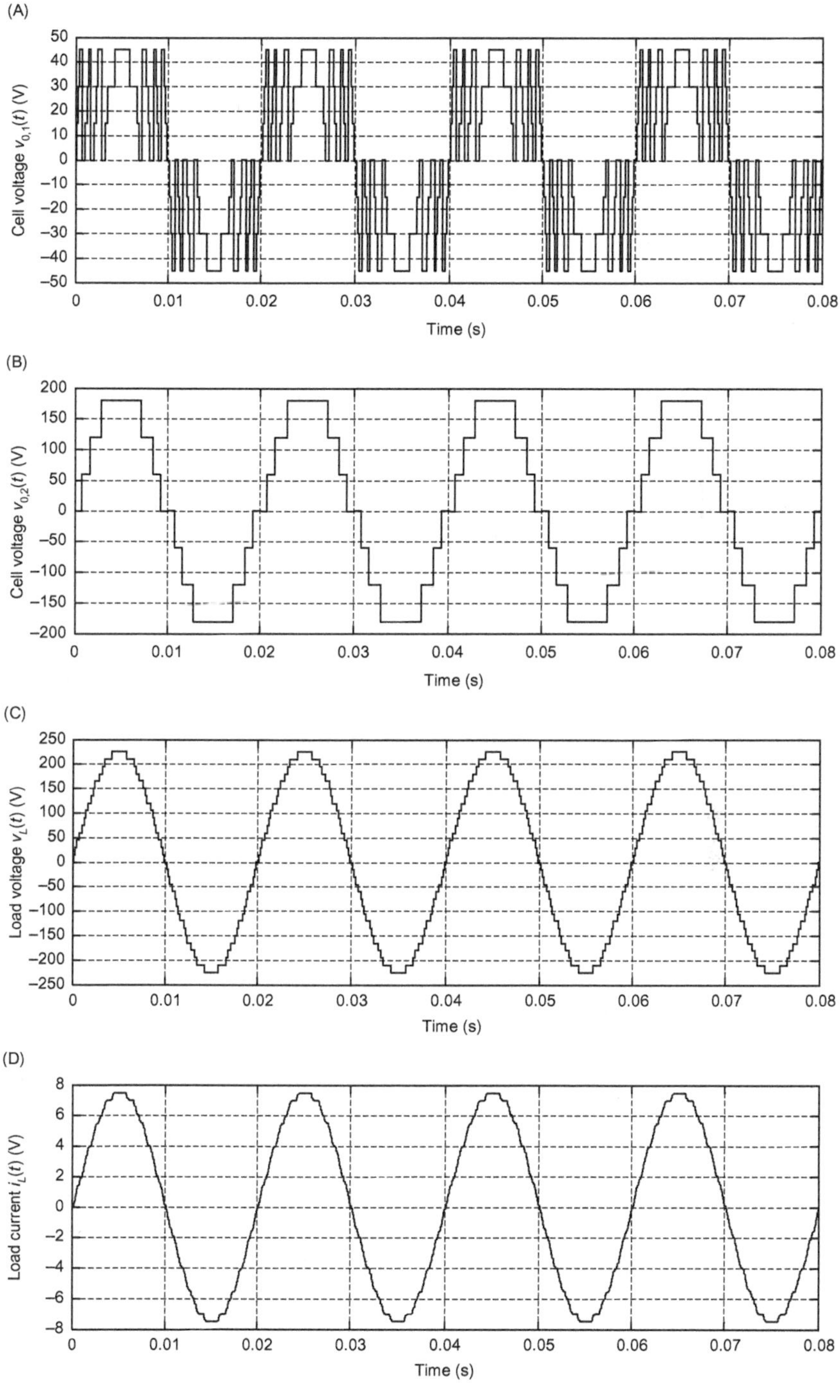

FIGURE 6.34 Waveforms for 31-level cascaded inverter (A) output voltage of cell 1 ($v_{0,1}(t)$); (B) output voltage of cell 2 ($v_{0,2}(t)$); (C) load voltage ($v_L(t)$); and (D) load current ($i_L(t)$).

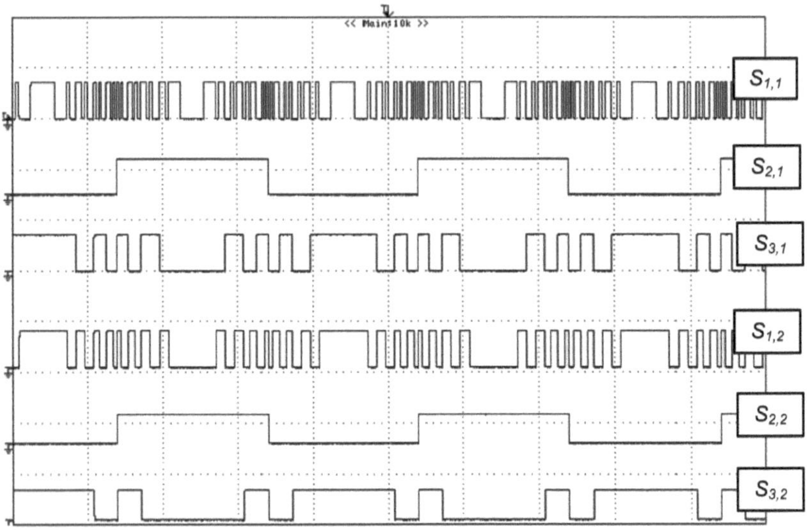

FIGURE 6.35 Experimentally obtained switching signals of the 31-level cascaded inverter (scale on y-axis: 5 V/div; scale on x-axis: 5 ms/div).

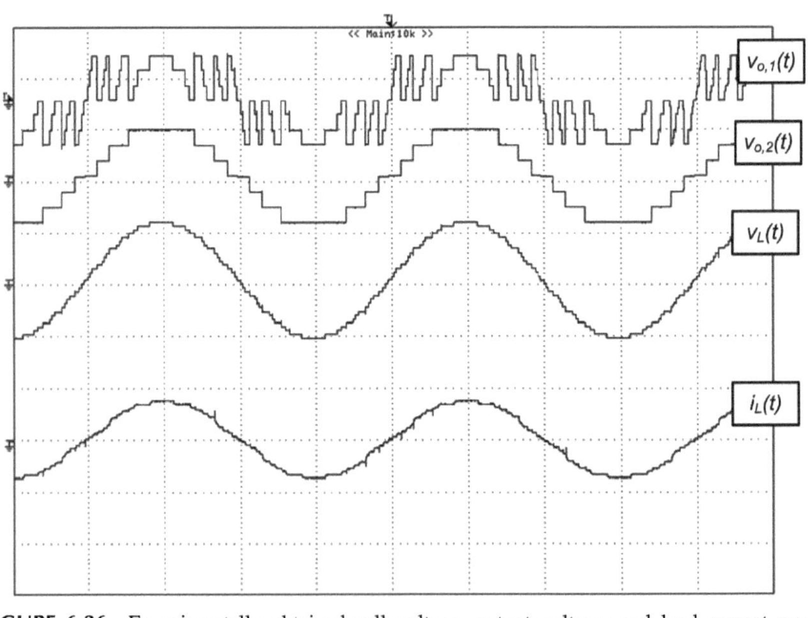

FIGURE 6.36 Experimentally obtained cell voltage, output voltage, and load current waveforms of the cascaded 31-level inverter ($v_{o,1}(t)$: 50 V/div; $v_{o,2}(t)$: 200 V/div; $v_L(t)$: 200 V/div; $i_L(t)$: 2 A/div; scale on x-axis: 5 ms/div).

6.9 SUMMARY

A novel topology for multilevel inverters is discussed in this chapter so as to reduce the device count as compared with the conventional topologies. The topology is obtained from the previously discussed BCS-MLI to overcome its limitations. This so-called CCS-MLI topology comprises floating input DC sources, alternately connected in opposite polarities with one another through power switches. All power switches are bidirectional-conducting-unidirectional-blocking, unlike the power switch requirement for BCS-MLI. Each input DC level appears in the stepped load voltage either individually or in additive combinations with other input levels. The working of the topology is explained and mathematical formulations corresponding to output voltage, source currents, and voltage stresses on switches are developed. Simulation and experimental studies performed on a five-level inverter based on the CCS-MLI structure were also discussed in this chapter. CCS-MLI can be effectively employed for applications where isolated DC sources are available. Investigations are carried out to explore power balancing and optimal switching for CCS-MLI having more than two equal-input DC sources. In addition, an algorithm for asymmetric source configuration is also presented since the commonly used asymmetric source configurations cannot be employed on CCS-MLI when an output voltage waveform with equal steps is desired.

In the second part of this chapter, the concept of cascaded multilevel inverter using submultilevel cells is discussed. Using the cascaded structure having CCS-MLI as the cells instead of generalized CCS-MLI offers important advantages. The cascaded topology exhibits the following characteristics:

1. Requirement of bidirectional-conducting-unidirectional-blocking switches only as compared to bidirectional-conducting-bidirectional-blocking switches;
2. Satisfactory applicability of symmetric source configuration;
3. Lesser voltage stress on the switches as compared to the operating voltage, especially for a large number of levels;
4. Low switching frequency (preferably line frequency) operation of higher voltage-rated switches;
5. Possibility of equal load sharing amongst all symmetric sources;
6. Improved modularity.

Connecting various submultilevel cells in series to obtain a cascaded multilevel inverter is an important approach to obtain a high-resolution voltage output and modularity of the apparatus. A cascaded structure using a CCS-MLI cell with two input sources is described in this chapter. Such a cell requires only bidirectional-conducting-unidirectional-blocking switches, while allowing fundamental frequency switching operation of higher voltage-rated switches. In addition, for symmetric sources, equal load sharing amongst the input sources is always permitted by the structure. Moreover, it allows asymmetric source configuration with binary source configuration. Studies with cascaded inverters with two input cells and the cases of symmetric and asymmetric source configurations were discussed in this chapter with nine-level and 31-level output voltage waveforms, respectively. Appropriate modulation strategies were also discussed.

REFERENCES

[1] S. Rohner, S. Bernet, M. Hiller, R. Sommer, Modulation, losses, and semiconductor requirements of modular multilevel converters, Ind. Electronics IEEE Trans. 57 (8) (2010) 2633–2642.

[2] C.-C. Hua, C.-W. Wu, C.-W Chuang, A novel dc voltage charge balance control for cascaded inverters, Power Electronics IET 2 (2) (2009) 147–155.

[3] J. Rodriguez, L.G. Franquelo, S. Kouro, J.I. Leon, R.C. Portillo, M.A.M. Prats, et al., Multilevel converters: an enabling technology for high-power applications, Proc. IEEE 97 (11) (2009) 1786–1817.

[4] E. Babaei, S.H. Hosseini, Charge balance control methods for asymmetrical cascade multilevel converters. Electrical Machines and Systems, 2007. ICEMS. International Conference on, vol., no., 8–11 Oct. 2007, pp. 74–79.

[5] X. Kou, K.A. Corzine, Y.L. Familiant, Full binary combination schema for floating voltage source multilevel inverters, Power Electronics IEEE Trans 17 (6) (2002) 891–897.

[6] E. Babaei, A cascade multilevel converter topology with reduced number of switches, Power Electronics IEEE Trans 23 (6) (2008) 2657–2664.

[7] J. Ebrahimi, E. Babaei, G.B Gharehpetian, A new multilevel converter topology with reduced number of power electronic components, Ind Electronics IEEE Trans 59 (2) (2012) 655–667.

[8] M.F. Kangarlu, E Babaei, A generalized cascaded multilevel inverter using series connection of submultilevel inverters, Power Electronics IEEE Trans 28 (2) (2013) 625–636.

[9] N.A. Rahim, M.M. Elias, W. Hew, Transistor-clamped h-bridge based cascaded multilevel inverter with new method of capacitor voltage balancing, Ind Electronics IEEE Trans 60 (8) (2013) 2943–2956.

[10] R. Naderi, A. Rahmati, Phase-shifted carrier pwm technique for general cascaded inverters, Power Electronics IEEE Trans 23 (3) (2008) 1257–1269.

[11] E. Solas, G. Abad, J. Barrena, S. Aurtenechea, A. Carcar, L. Zajac, Modular multilevel converter with different submodule concepts – Part I: capacitor voltage balancing method, Ind Electronics IEEE Trans 60 (10) (2013) 4525–4535.

[12] J. Pereda, J. Dixon, Cascaded multilevel converters: Optimal asymmetries and floating capacitor control, Ind Electronics IEEE Trans 60 (11) (2013) 4784–4793.

[13] T.Wanjekeche,D.V.Nicolae, A.A. Jimoh, A cascaded NPC/H-Bridge inverter with simplified control strategy and minimum component count. AFRICON, 2009. AFRICON '09., vol., no., 23–25 Sept. 2009, pp. 1–6.

[14] J. Ebrahimi, E. Babaei, G.B. Gharehpetian, A new topology of cascaded multilevel converters with reduced number of components for high-voltage applications, Power Electronics IEEE Trans 26 (11) (2011) 3109–3118.

[15] E. Babaei, S.H. Hosseini, G.B. Gharehpetian, H.M. Tarafdar, M Sabahi, Reduction of dc voltage sources and switches in asymmetrical multilevel converters using a novel topology, Electric Power Syst Res 77 (8) (2007) 1073–1085.

[16] E. Babaei, S.H. Hosseini, New cascaded multilevel inverter topology with minimum number of switches, Energy Convers Manage 50 (11) (2009) 2761–2767.

[17] M. Nilkar, E. Babaei, M. Sabahi, A new single-phase cascade multilevel inverter topology using four-level cells, Electrical Engineering (ICEE), 2012 20th Iranian Conference on, vol., no., 15–17 May 2012, pp. 348–353.

[18] M.F. Kangarlu, E. Babaei, S. Laali, Symmetric multilevel inverter with reduced components based on non-insulated dc voltage sources, Power Electronics IET 5 (5) (2012) 571–581.

[19] H. Akagi, Classification, terminology, and application of the modular multilevel cascade converter (MMCC), Power Electronics IEEE Trans. 26 (11) (2011) 3119–3130.

[20] C.G. Hochgraf, High voltage modular inverter. U.S. Patent 6072707, 06 June, 2000.

[21] A. Larrinaga, L. Bengoetxea, E. Llanos, Modular converter based on multi-level distributed circuits with a capacitive mid-point. World Patent WO2011073466A1, 23 June 2011.

Chapter 7

Comparison of Multilevel Inverter Topologies

7.1 INTRODUCTION

Comparing various multilevel topologies is a difficult task because each topology is intrinsically different and has its own distinct features characterized by power circuits, control flexibility, power switch requirements, number of switches and total voltage-blocking requirement, source configuration, modularity, etc. To draw a common framework for comparison of various topologies may not just be futile, but also unfair. The employability of a given topology is mainly settled by its application requirements.

However, because we have mainly discussed device count reduction in various topologies, a comparison amongst them is carried out here primarily on the basis of the number of power switches present in the power circuit. The number of power switches and their voltage rating requirements give an idea about the reliability and losses (conduction and switching) of the respective inverter. The number of power switches also determines the volume added to the inverter with gate driver units and protection circuits. Total voltage blocking (many researchers call it "total standing voltage" [TSV]) for power switches is an important criterion to evaluate the characteristics of an inverter [1,2]. It gives an idea of the total amount of semiconductor used in the circuit and partly reflects the costs and losses incurred.

In this chapter a paradigm is presented for qualitative and quantitative comparison of topologies. That is, once you have a topology at hand, how do you compare it with others? This is shown with the examples of comparisons of topologies discussed in this book with the existing topologies. In doing so, graphical plots are used to help you visualize the comparison. While comparing the topologies, only the symmetric source configuration is considered. This is because there are many asymmetric configurations and a topology may not support all of them (sometimes not even one of them). Hence, those topologies which require compulsory asymmetric source configuration are left out of the quantitative comparison. This means that the "packed U-cell topology" discussed in Chapter 3, Advent of New Topologies, and BCS-MLI discussed in Chapter 5, Multilevel Inverter Based on Bridge-Type Connected Sources, do not appear in this quantitative comparison of topologies.

Multilevel Inverters. DOI: https://doi.org/10.1016/B978-0-12-812448-2.00007-6

In the context of the discussions in previous chapters, the general criteria for an overall assessment of a topology are formulated as:

1. The number of power switches used, irrespective of the voltage rating of the switches;
2. The total blocking voltage requirement for the converter (which depends on the number of power switches and their respective voltage ratings).

Other qualitative issues, such as controllability of the topologies in terms of possibility of charge balance control, appropriate switching of different voltage-rated switches, and possibility of employing asymmetric sources, have been commented upon in earlier chapters.

First, two topologies, namely, CCS-MLI (described in Chapter 6: Cross-Connected Sources-Based Multilevel Inverter) and cascaded MLI (also described in Chapter 6: Cross-Connected Sources-Based Multilevel Inverter) are compared with the neutral point clamped (NPC), cascaded H-bridge (CHB), and flying capacitor (FC) topologies, as they are considered to be the standard in multilevel topologies. Subsequently, these two topologies are compared with the advanced topologies discussed in Chapter 3, Advent of New Topologies (excluding the packed U-cell topology).

7.1.1 Comparison With Classical Topologies

The requirements of various components for the CCS-MLI and cascaded MLI and the so-called classical topologies are shown in Table 7.1 for symmetric source configuration and three-phase implementation. It is important to note that individual switches and diodes are considered independent of their voltage-blocking capability. A plot of the number of levels versus the component count is shown in Fig. 7.1 and indicates that CCS-MLI and cascaded MLI substantially reduce the component count, especially for a large number of levels. For an 11-level waveform in the phase voltage, for example, the NPC, FC, and CHB topologies require 184, 158, and 135 components, respectively, whereas the CCS-MLI topology requires 87 components and the cascaded MLI requires 105 components. When the power switch requirement for an 11-level voltage waveform is considered, the classical topologies require 60 power switches each, whereas the CCS-MLI and cascaded MLI topologies require 36 and 45 switches, respectively. A comparison of these topologies on the basis of power switch requirement is also shown in Fig. 7.2.

It can be observed from Table 7.1 that although the number of power switches and the component count are reduced with CCS-MLI and cascaded MLI, the total blocking voltage remains the same. Many topologies, as discussed in Chapter 3, Advent of New Topologies, reduce the power switch count and require fewer switches, but are of very high blocking capability, thereby increasing the total blocking voltage. Total voltage blocking for

TABLE 7.1 Comparison of Three-Phase Multilevel Inverters (Classical Topologies and CCS-MLI and Cascaded MLI Topologies) (n_{levels} = Number of Levels in Phase Voltage)

Inverter Type/Component	Classical Topologies			Topologies Discussed in Chapter 6, Cross-Connected Sources-Based Multilevel Inverter	
	NPC	FC	CHB	CCS-MLI	Cascaded Inverter $N = 5,9,13,17,\ldots(4j+1)$ $\{\text{for } j = 1,2,3,\ldots\}$
Number of main switches	$6(n_{levels}-1)$	$6(n_{levels}-1)$	$6(n_{levels}-1)$	$3(n_{levels}+1)$	$\frac{9(n_{levels}-1)}{2}$
Number of main diodes	$6(n_{levels}-1)$	$6(n_{levels}-1)$	$6(n_{levels}-1)$	$3(n_{levels}+1)$	$\frac{9(n_{levels}-1)}{2}$
Number of clamping diodes	$6(n_{levels}-2)$	0	0	0	0
Number of DC bus capacitors/Isolated supplies	$(n_{levels}-1)$	$(n_{levels}-1)$	$\frac{3(n_{levels}-1)}{2}$	$\frac{3(n_{levels}-1)}{2}$	$\frac{3(n_{levels}-1)}{2}$
Number of flying capacitors	0	$3(n_{levels}-2)$	0	0	0
Total component count	*19* $(n_{levels}-25)$	$2(8n_{levels}-9)$	$\frac{27(n_{levels}-1)}{2}$	$\frac{3(5n_{levels}+3)}{2}$	$\frac{21(n_{levels}-1)}{2}$
Total blocking voltage for main switches (in p.u.)	$6(n_{levels}-1)$	$6(n_{levels}-1)$	$6(n_{levels}-1)$	$6(n_{levels}-1)$	$6(n_{levels}-1)$
Total blocking voltage for main diodes (in p.u.)	$6(n_{levels}-1)$	$6(n_{levels}-1)$	$6(n_{levels}-1)$	$6(n_{levels}-1)$	$6(n_{levels}-1)$
Total voltage stress borne by DC bus capacitors/Isolated supplies (in p.u.)	$3(n_{levels}-1)$	$3(n_{levels}-1)$	$\frac{3(n_{levels}-1)}{2}$	$\frac{3(n_{levels}-1)}{2}$	$\frac{3(n_{levels}-1)}{2}$

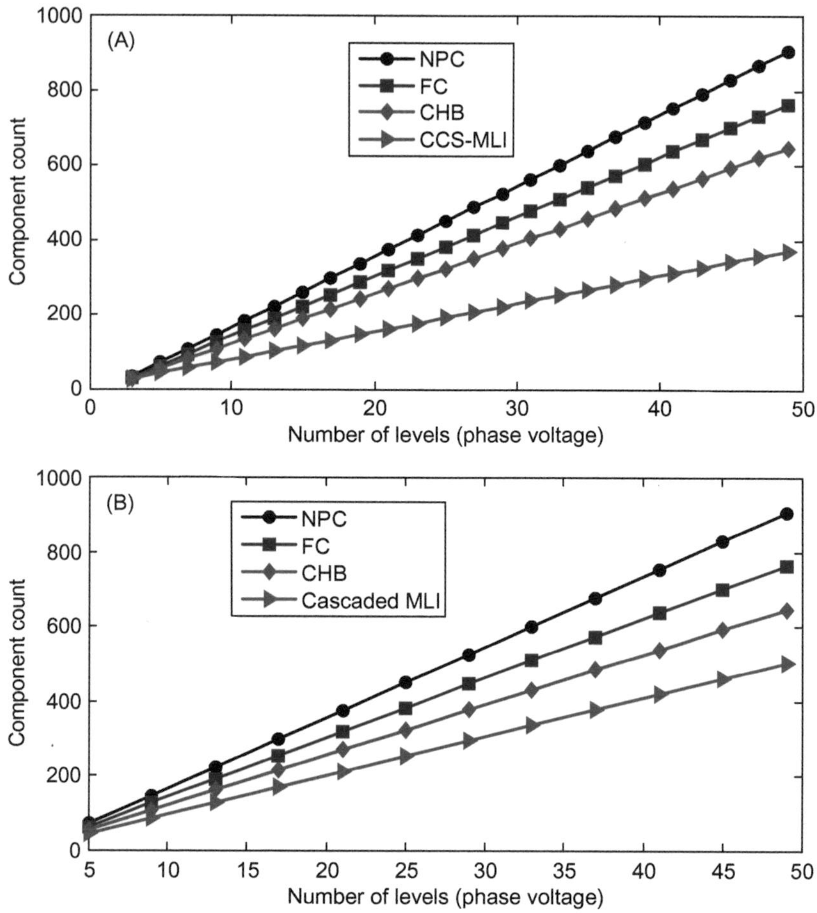

FIGURE 7.1 Comparison of MLI topologies for three-phase implementation on the basis of component count: (A) CCS-MLI versus classical topologies; and (B) cascaded MLI versus classical topologies.

power switches is an important criterion to evaluate the characteristics of an inverter [1,2]. It gives an idea of the total amount of semiconductor used in the circuit and partly reflects the costs and losses incurred.

7.1.2 Comparison With Reduced Device Count (RDC) Topologies

In Chapter 3, Advent of New Topologies, nine topologies were discussed which were proposed by the respective researchers with the objective of reducing the component count for an increased number of output levels. These topologies are referred to as reduced device count multilevel inverters (RDC-MLIs). In this

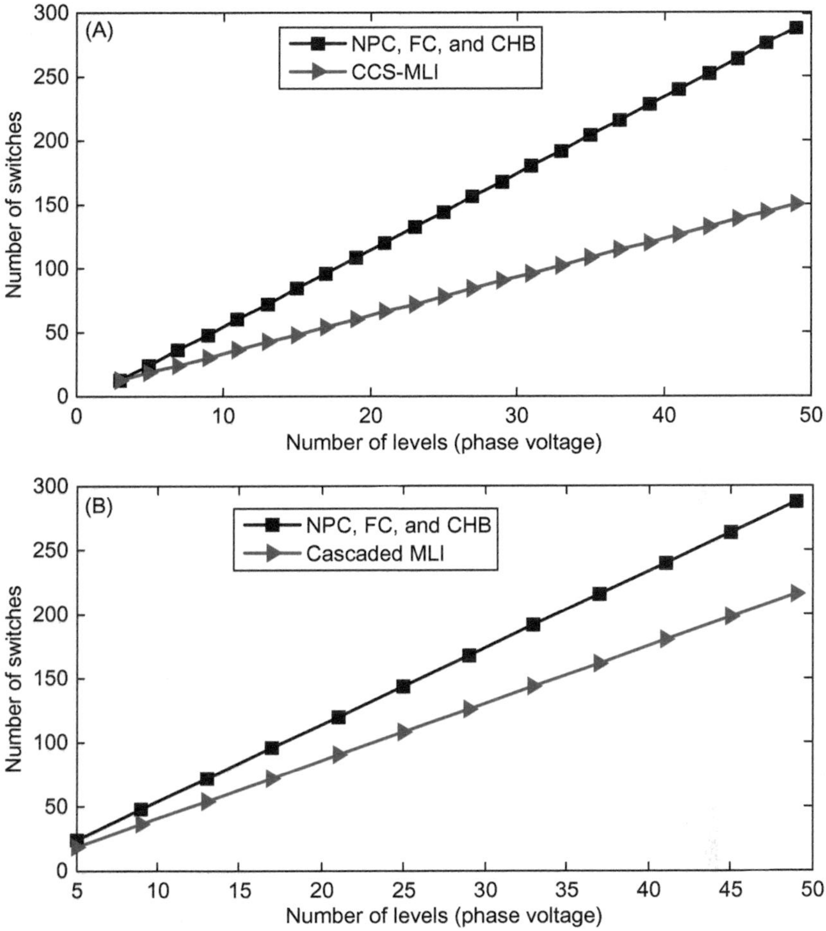

FIGURE 7.2 Comparison of MLI topologies for three-phase implementation on the basis of power switch requirement: (A) CCS-MLI versus classical topologies; and (B) cascaded MLI versus classical topologies.

subsection a comparison of CCS-MLI and cascaded MLI proposed in this work is carried out with these topologies (excluding PUC). For three-phase implementation, various requirements are summarized in Table 7.2.

As shown in Fig. 7.3, CCS-MLI not only requires a lower number of switches as compared to "topology #1 (MLDCL inverter)," the total voltage-blocking capability required is also lower. Both topologies require isolated DC sources. However, unlike CCS-MLI, "topology #1" always facilitates equal load sharing amongst the input DC sources.

"Topology #2 (T-type inverter)" not only requires a larger number of power switches than CCS-MLI, the total blocking voltage of switches is also

TABLE 7.2 Comparison of Three-Phase Multilevel Inverters (n_{levels} = Number of Levels in Phase Voltage)

Topology #	Topology	Number of BCUB Switches	Number of BCBB Switches	Total Voltage Stress on Switches (p.u.)	Isolated Sources/ DC Capacitors	Total Components
1	MLDCL inverter (refer to Section 3.3.1)	$3(n_{levels}+3)$	0	$9[(n_{levels}-1)]$	$\frac{3(n_{levels}-1)}{2}$	$\frac{3(3n_{levels}+5)}{2}$
2	T-type inverter (refer to Section 3.3.2)	12	$3(n_{levels}-1)$	$3[2(n_{levels}-1)+\sum_{k=1}^{(n_{levels}-3)/4}(n_{levels}-(2k+1))]$, when number of sources is odd $3\left[\frac{9(n_{levels}-1)}{4}+\sum_{k=1}^{(n_{levels}-5)/4}(n_{levels}-(2k+1))\right]$, when number of sources is even	$\frac{3(n_{levels}-1)}{2}$	$\frac{3(3n_{levels}-15)}{2}$
3	SSPS-MLI (refer to Section 3.3.3)	$\frac{3(3n_{levels}-1)}{2}$	0	$\frac{3[(7n_{levels}-13)]}{2}$	$\frac{3(n_{levels}-1)}{2}$	$3(2n_{levels}-1)$
4	SCSS-MLI (refer to Section 3.3.4)	$3(n_{levels}+3)$	0	$3\left[(3n_{levels}-2)+\sum_{k=2}^{(n_{levels}-3)/2}k\right]$	$\frac{3(n_{levels}-1)}{2}$	$\frac{3(n_{levels}+5)}{2}$

5	CBSC-MLI (refer to Section 3.3.5)	0	$3(n_{levels}+1)$	$3\left[\frac{(n_{levels}+1)(3n_{levels}-1)}{8}\right]$, when number of sources is odd $3\left[\frac{(n_{levels}-1)(3n_{levels}+5)}{8}\right]$, When number of sources is even	$\frac{3(n_{levels}-1)}{2}$	$\frac{3(n_{levels}+1)}{2}$
7	MLM-MLI (refer to Section 3.3.7)	12	$\frac{3(n_{levels}+1)}{2}$	$3\left[3(n_{levels}-1)+\sum_{k=1}^{(n_{levels}-3)/4}\{n_{levels}-(2k+1)\}\right]$, when number of sources is odd $3\left[\frac{13(n_{levels}-1)}{4}+\sum_{k=1}^{(n_{levels}-5)/4}\{n_{levels}-(2k+1)\}\right]$, when number of sources is even	$\frac{3(n_{levels}-1)}{2}$	$3(n_{levels}+4)$
8	RV topology (refer to Section 3.3.8)	$3(n_{levels}+3)$	0	$9(n_{levels}-1)$	$\frac{(n_{levels}-1)}{2}$	$\frac{(7n_{levels}+17)}{2}$
9	2SELG-MLI (refer to Section 3.3.9)	24	$\frac{3(n_{levels}-7)}{2}$	$V_o + 4n;\ n=\frac{(n_{levels}-1)}{2}$, where, $V_o=\begin{cases}\frac{3n^2+8n+5}{5} \text{ when } n \text{ is odd and } \frac{n-1}{2} \text{ is even}\\ \frac{3n^2+8n+5}{5} \text{ when } n \text{ is odd and } \frac{n-1}{2} \text{ is odd}\\ \frac{3n^2+8n+1}{5} \text{ when } n \text{ is even and } \frac{n}{2} \text{ is odd}\\ \frac{3n^2+8n}{5} \text{ when } n \text{ is even and } \frac{n}{2} \text{ is even}\end{cases}$	$\frac{3(n_{levels}-1)}{2}$	$3(n_{levels}+4)$

BCUB, bidirectional-conducting-unidirectional-blocking; *BCBB*, bidirectional-conducting-bidirectional-blocking.

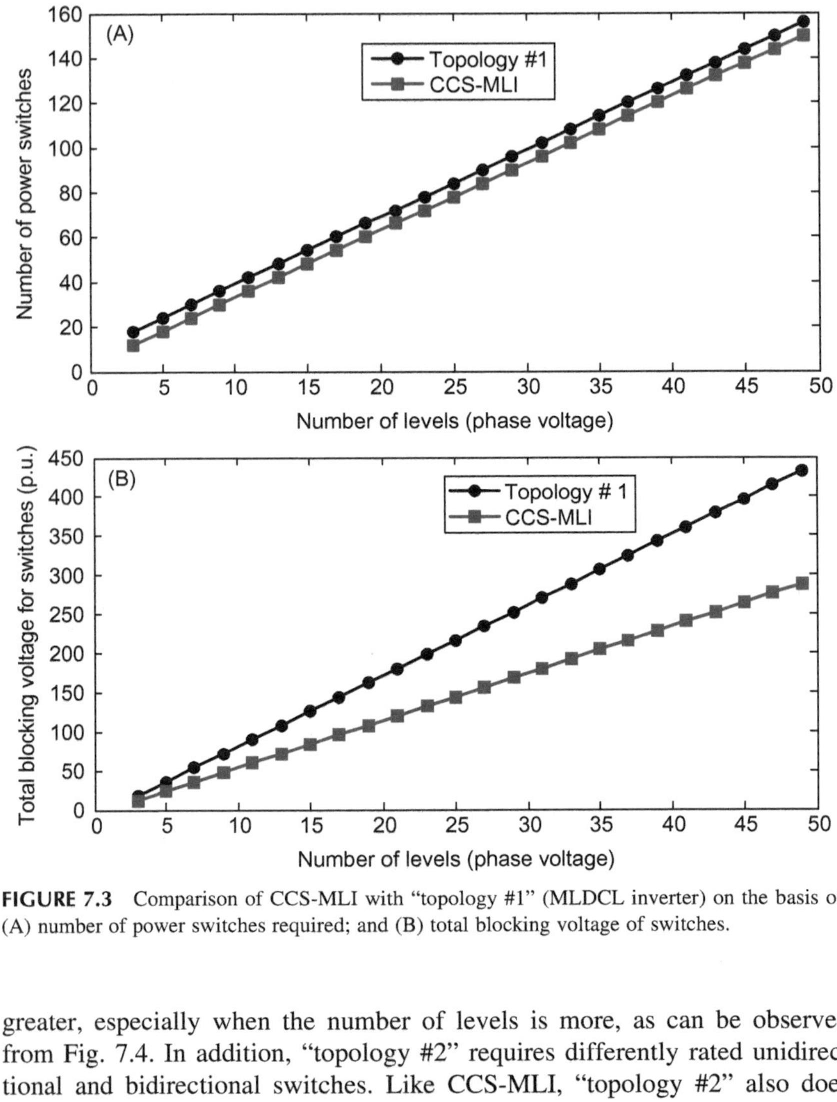

FIGURE 7.3 Comparison of CCS-MLI with "topology #1" (MLDCL inverter) on the basis of: (A) number of power switches required; and (B) total blocking voltage of switches.

greater, especially when the number of levels is more, as can be observed from Fig. 7.4. In addition, "topology #2" requires differently rated unidirectional and bidirectional switches. Like CCS-MLI, "topology #2" also does not allow equal load sharing amongst input DC sources. However, unlike CCS-MLI, "topology #2" requires nonisolated input DC levels.

A comparison between CCS-MLI and "topology #3" (SSPS-MLI) is shown in Fig. 7.5 on the basis of power switch requirement and total blocking voltage of switches. CCS-MLI requires a significantly lower number of power switches while keeping the total blocking voltage for switches lower. However, unlike CCS-MLI, "topology #3" allows equal load sharing amongst the input DC sources.

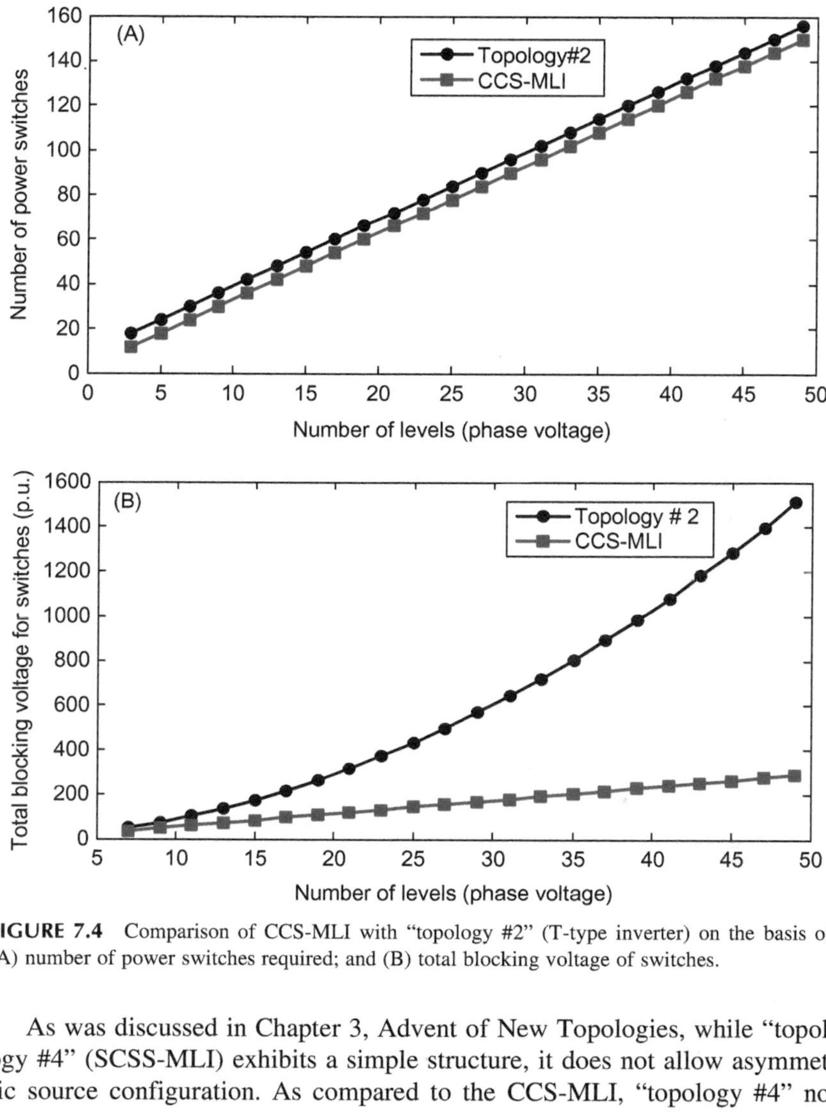

FIGURE 7.4 Comparison of CCS-MLI with "topology #2" (T-type inverter) on the basis of: (A) number of power switches required; and (B) total blocking voltage of switches.

As was discussed in Chapter 3, Advent of New Topologies, while "topology #4" (SCSS-MLI) exhibits a simple structure, it does not allow asymmetric source configuration. As compared to the CCS-MLI, "topology #4" not only requires a greater number of switches, but the total blocking voltage is also higher, as can be observed from Fig. 7.6. However, "topology #4," allows fundamental frequency switching of the highest voltage-rated switches.

In terms of number, the requirement for switches is the same in "topology #5" (CBSC-MLI) as that of CCS-MLI. All switches, however, are bidirectional-conducting-bidirectional-blocking in "topology #5," and are differently voltage-rated. As can be observed from Fig. 7.7, the total blocking

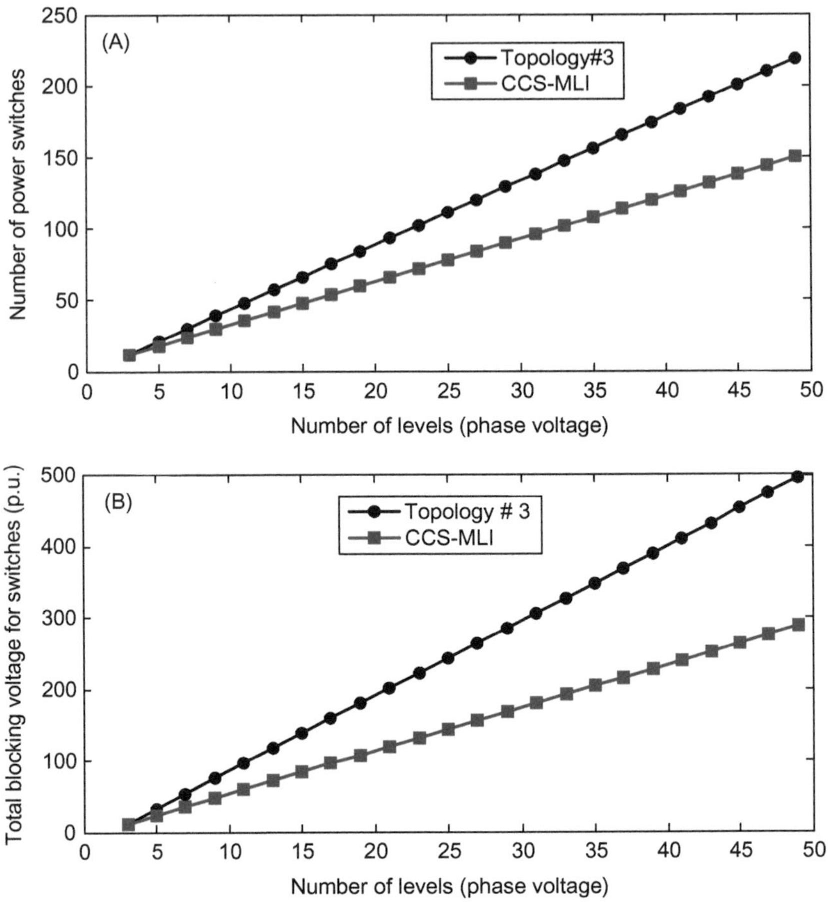

FIGURE 7.5 Comparison of CCS-MLI with "topology #3" (SSPS-MLI) on the basis of: (A) number of power switches required; and (B) total blocking voltage of switches.

voltage for switches is significantly higher as compared to CCS-MLI, especially for a large number of levels in the output voltage.

As was discussed in Chapter 3, Advent of New Topologies, "topology #6" (PUC topology) does not support a symmetric source configuration and requires a rather complex algorithm for incorporating asymmetry in the source voltage ratios. The structure, however, is simple.

While comparing the CCS-MLI with "topology #7" (MLM-MLI), as shown in Fig. 7.8, it can be seen that the number of switches required in CCS-MLI is more, however, all the switches needed in "topology #7" are bidirectional-conducting-unidirectional-blocking except the four H-bridge

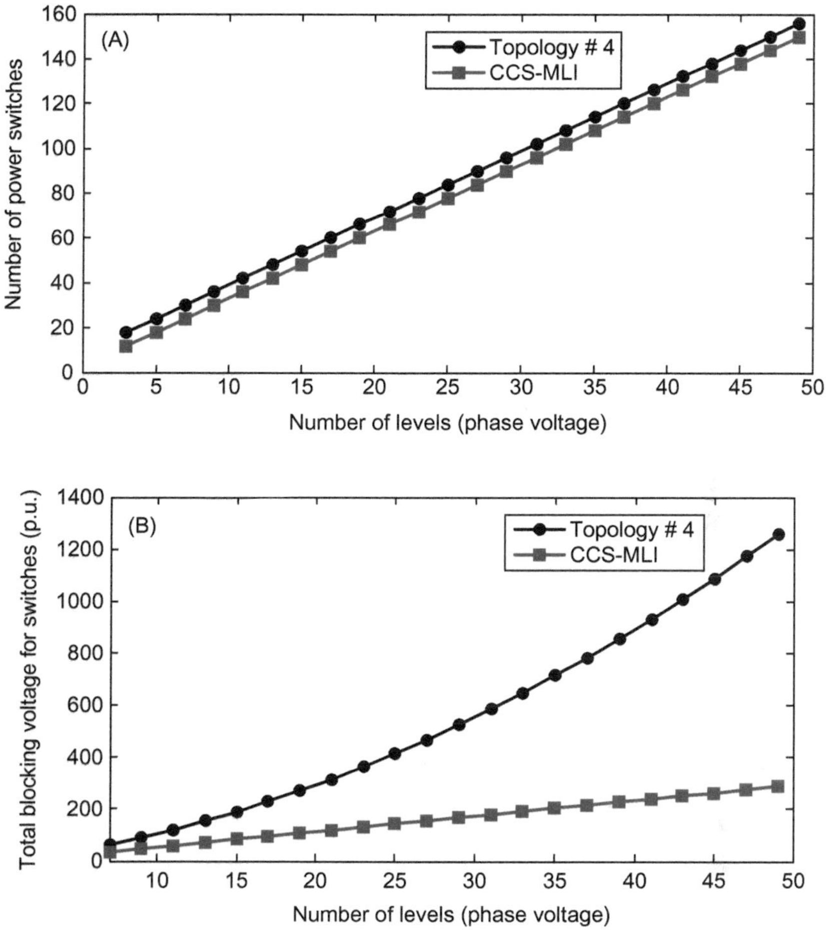

FIGURE 7.6 Comparison of CCS-MLI with "topology #4" (SCSS-MLI) on the basis of: (A) number of power switches required; and (B) total blocking voltage of switches.

switches. In addition, the total blocking voltage is much greater in the case of "topology #7." The topology requires nonisolated input levels, unlike the CCS-MLI. In addition, unlike the CCS-MLI, "topology #7" does not allow asymmetric source configuration.

"Topology #8" (RV topology) shows an important advantage over the CCS-MLI in terms of the requirement for input sources, as it requires a single link for three-phase implementation. Still, both in terms of power switch requirement and total blocking voltage for switches, the CCS-MLI is advantageous over "topology #8," as can be observed from Fig. 7.9.

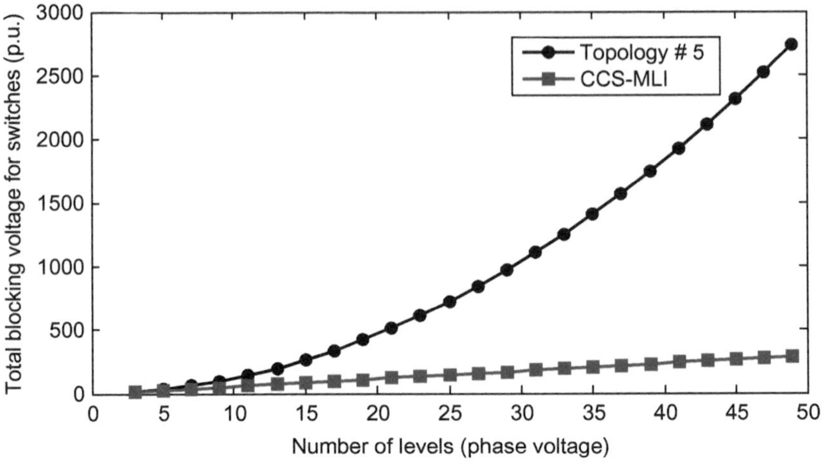

FIGURE 7.7 Comparison of CCS-MLI with "topology #5" (CBSC-MLI) on the basis of total blocking voltage of switches.

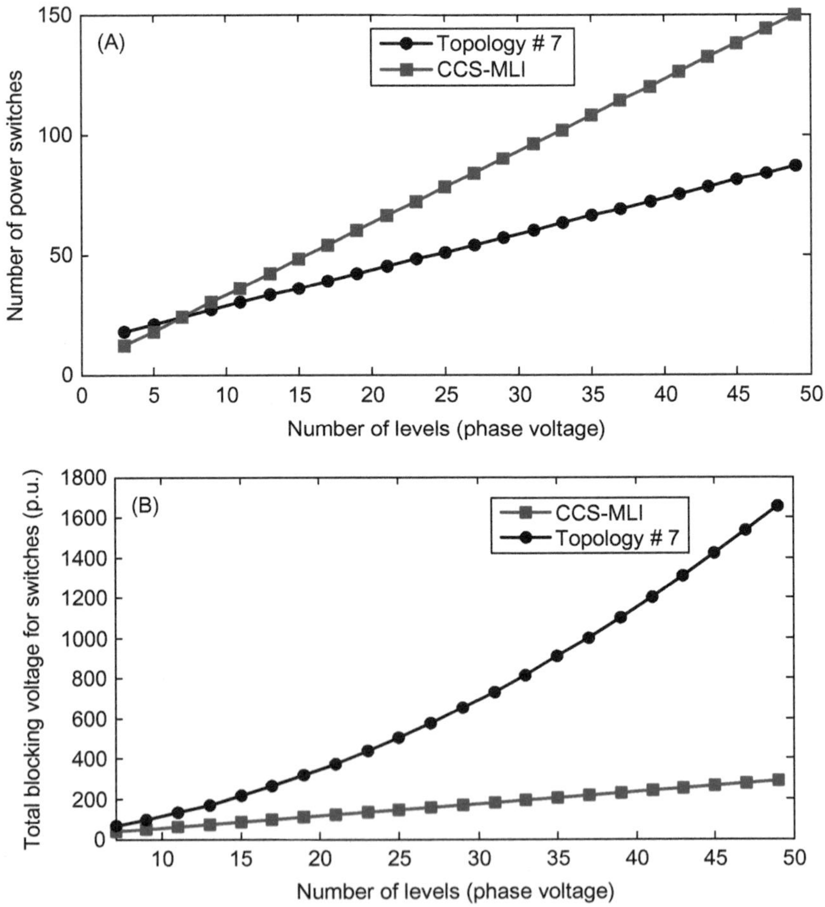

FIGURE 7.8 Comparison of CCS-MLI with "topology #7" (MLM-MLI) on the basis of: (A) number of power switches required; and (B) total blocking voltage of switches.

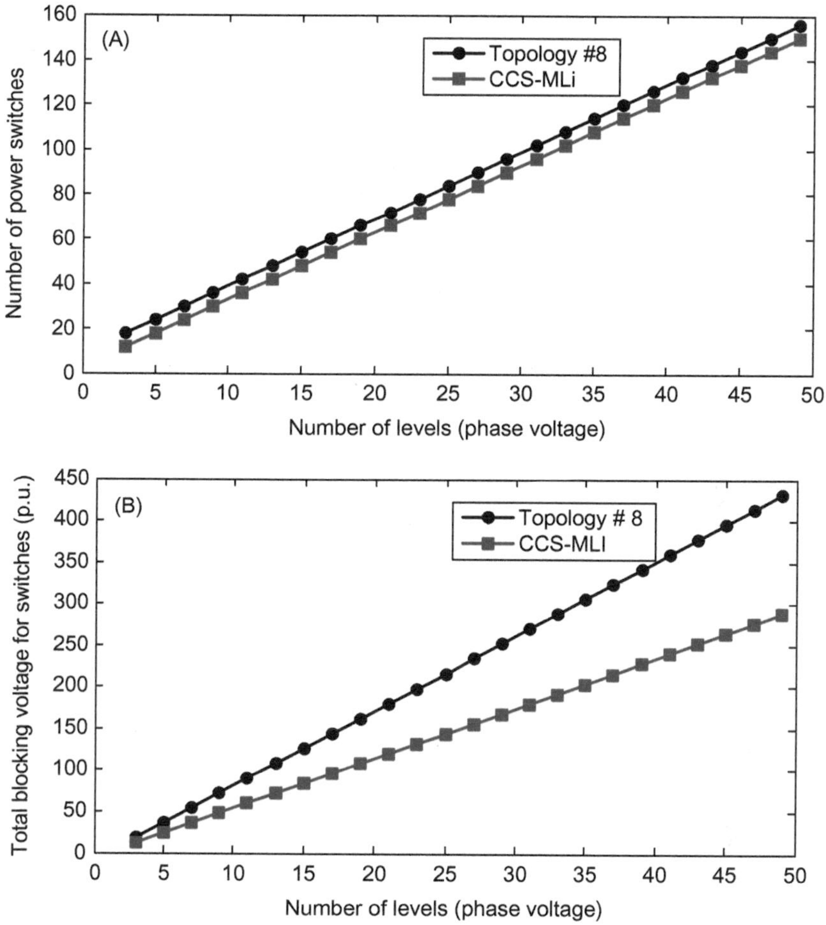

FIGURE 7.9 Comparison of CCS-MLI with "topology #8" (RV topology) on the basis of: (A) number of power switches required; and (B) total blocking voltage of switches.

Considering "topology #9" (2SELG-MLI), it requires a mix of bidirectional-conducting-bidirectional-blocking and bidirectional-conducting-unidirectional-blocking switches and, in terms of number of switches, it shows requirement for a lower number of power switches as compared to the CCS-MLI, as shown in Fig. 7.10. The CCS-MLI manifests lower total voltage-blocking capability for the switches for greater than 35 levels.

In Chapter 6, Cross-Connected Sources-Based Multilevel Inverter, an alternative generalization of CCS-MLI was also discussed to obtain a cascaded inverter so as to incorporate features such as modularity, equal load sharing amongst all input sources, and fundamental frequency switching of higher voltage-rated switches. It can be observed from Fig. 7.11 that, as

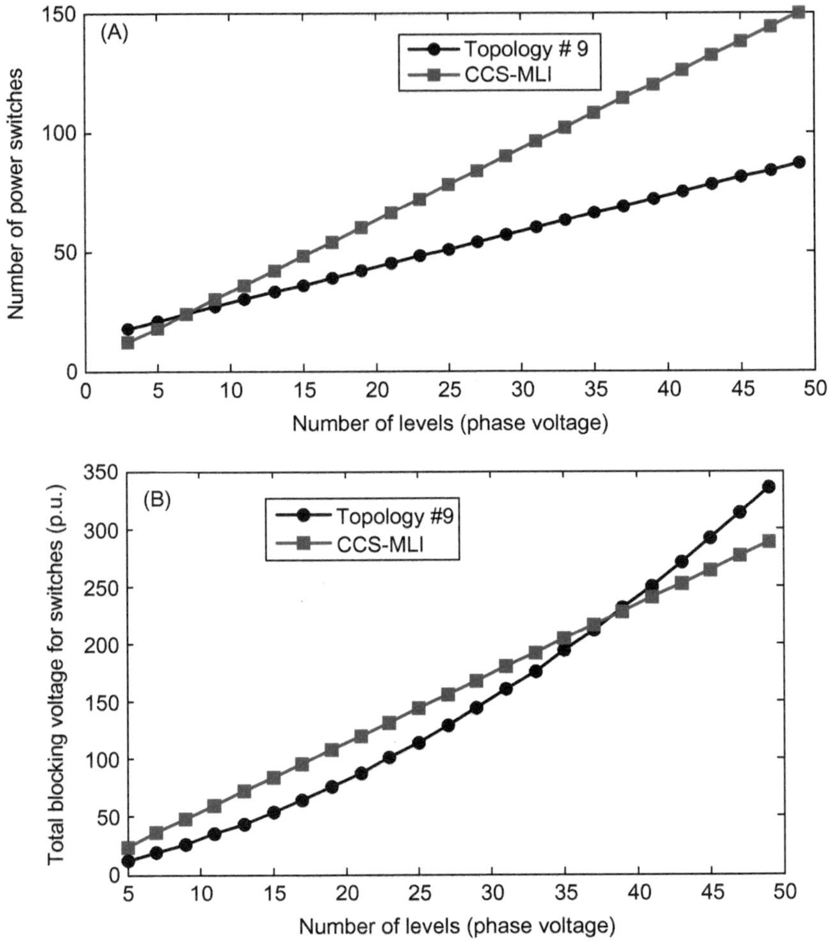

FIGURE 7.10 Comparison of CCS-MLI with "topology #9" (2SELG-MLI) on the basis of: (A) number of power switches required; and (A) total blocking voltage of switches.

compared to the classical topologies, the cascaded MLI requires less power switches while maintaining equal total voltage-blocking requirements.

In addition, incorporating desirable features leaves the cascaded multilevel inverter with a greater requirement for power switches and overall component count as compared to the original CCS-MLI, as can be observed from Fig. 7.12. The total blocking voltage requirement of the switches, however, remains the same. It can be noted that the cascaded inverter requires two input DC sources in each module. While the CCS-MLI requires a lower number of power switches as compared to those of many topologies under discussion, the cascaded MLI may not be as advantageous, as can be observed in Figs. 7.12–7.20. It can be especially noted that the cascaded

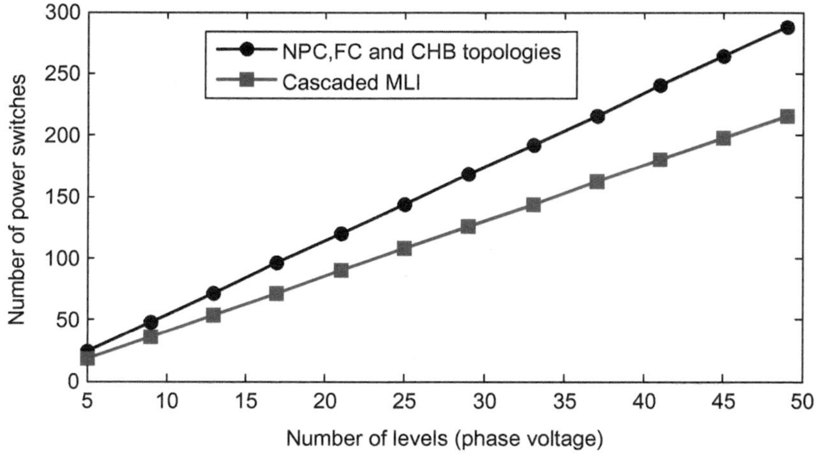

FIGURE 7.11 Comparison of cascaded MLI with classical topologies on the basis of number of power switches required.

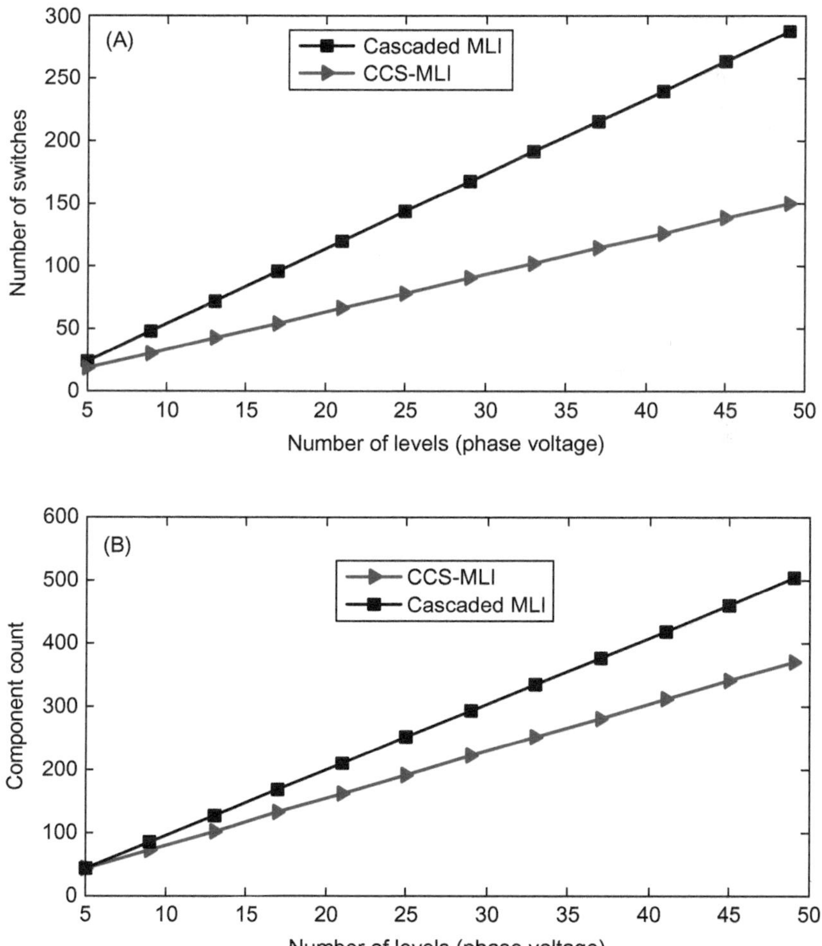

FIGURE 7.12 Comparison of cascaded MLI with CCS-MLI on the basis of: (A) number of power switches required; and (A) component count.

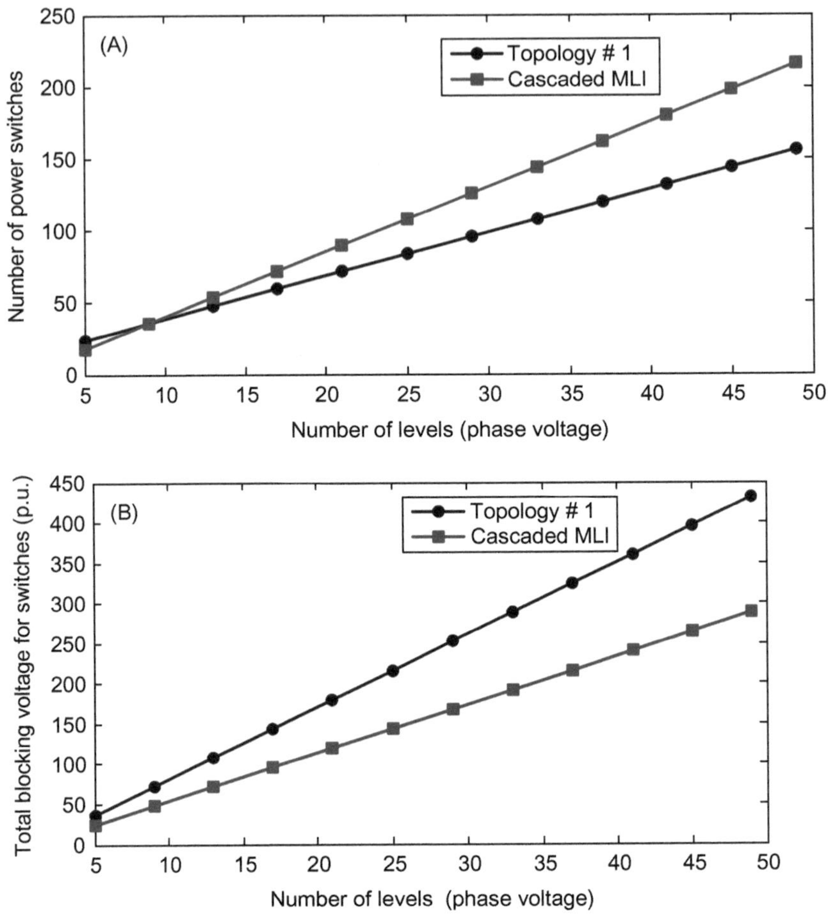

FIGURE 7.13 Comparison of cascaded MLI with "topology #1" (MLDCL inverter) on the basis of: (A) number of power switches required; and (B) total blocking voltage of switches.

inverter manifests a lower requirement for total voltage blocking for the power switches. In addition, on account of the benefits offered by the cascaded inverter, the compromise in terms of number of power switches may be worthwhile.

7.2 SUMMARY

The comparison of multilevel inverter topologies is a tricky issue on account of the distinct features of the different topologies. In this book, topologies

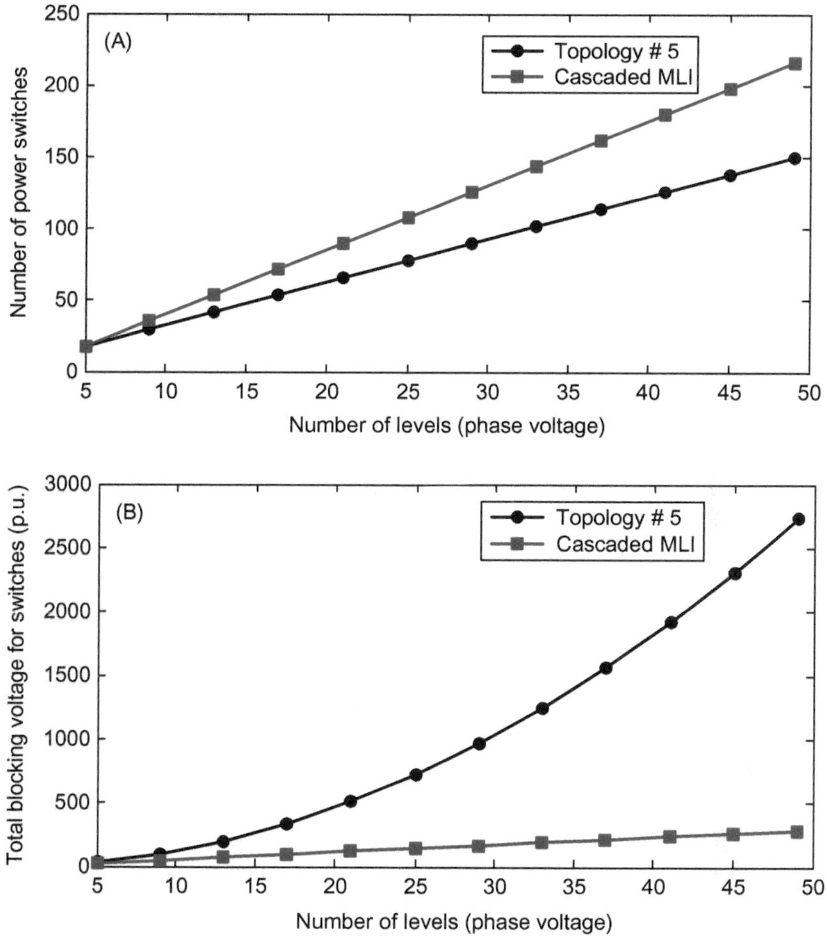

FIGURE 7.14 Comparison of cascaded MLI with "topology #2" (T-type inverter) on the basis of: (A) number of power switches required; and (B) total blocking voltage of switches.

with RDCs are discussed and in this chapter a comparison is presented between topologies primarily on the basis of power switch count and total voltage-blocking requirement. Topologies which require compulsory asymmetric source configuration have been excluded.

Based on quantitative analysis, it can be concluded that:

- The cross-connected sources-based multilevel inverter (CCS-MLI) topology requires a much lower number of power switches as compared to the classical topologies. In addition, the overall component count is also lower, while the total voltage-blocking capability of the switches is equal to that of the classical topologies. When compared to the other topologies with RDCs, CCS-MLI seems to be very competitive.

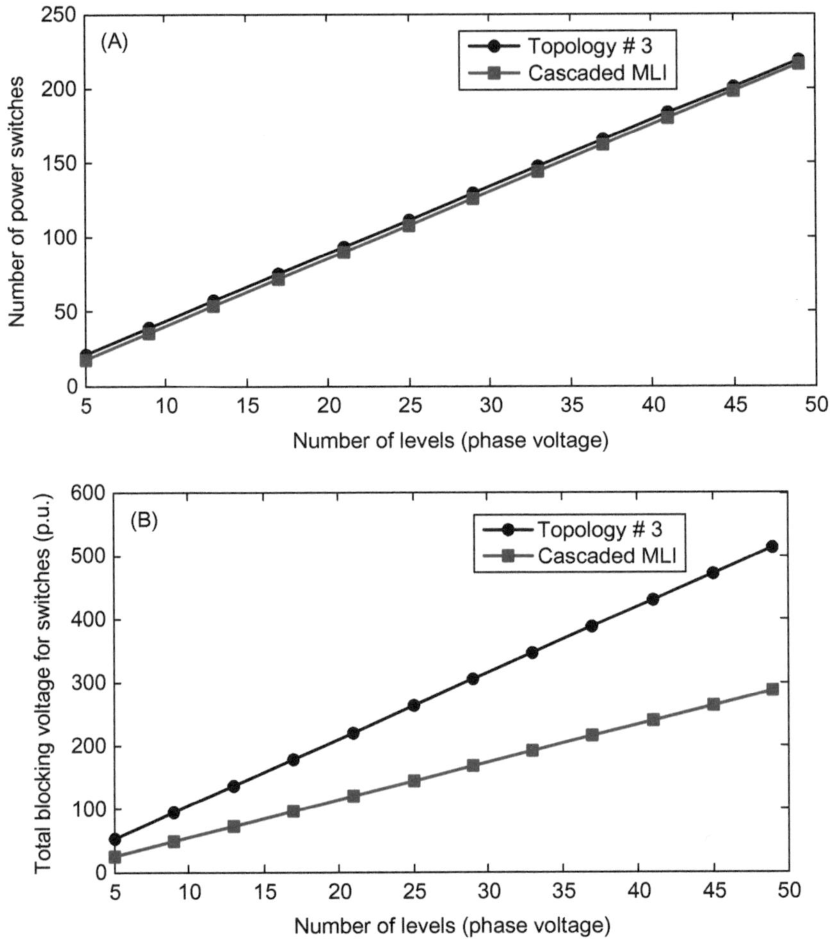

FIGURE 7.15 Comparison of cascaded MLI with "topology #3" (SSPS-MLI) on the basis of: (A) number of power switches required; and (B) total blocking voltage of switches.

- The cascaded topology based on CCS-MLI, however, requires slightly more power switches as it incorporates important practical features such as equal load sharing amongst all input sources and fundamental frequency switching of higher voltage-rated switches. More importantly, the total voltage-blocking requirement of the proposed cascaded inverter remains the same as that of the classical topologies. Thus both the proposed structures are fairly competent as compared to the existing structures.

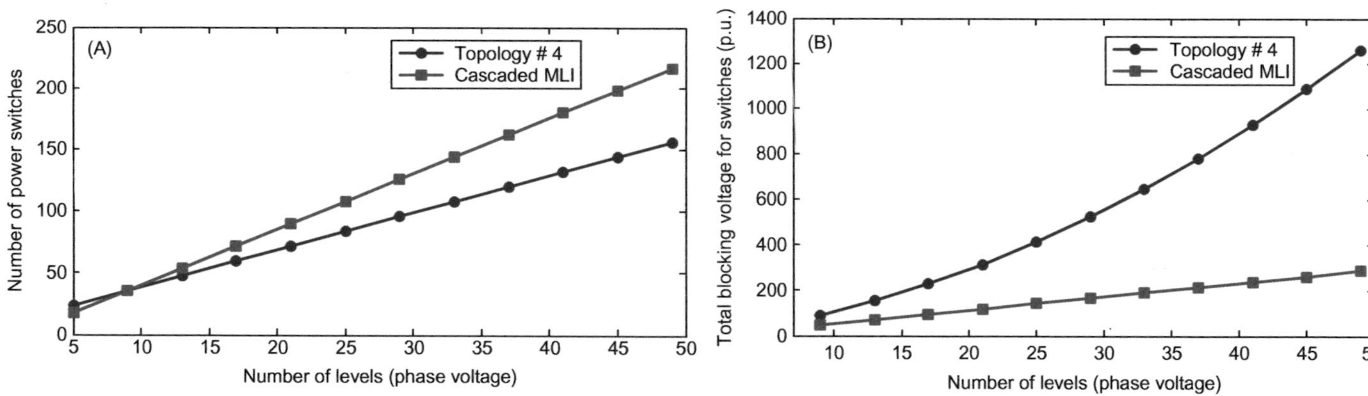

FIGURE 7.16 Comparison of cascaded MLI with "topology #4" (SCSS-MLI) on the basis of: (A) number of power switches required; and (B) total blocking voltage of switches.

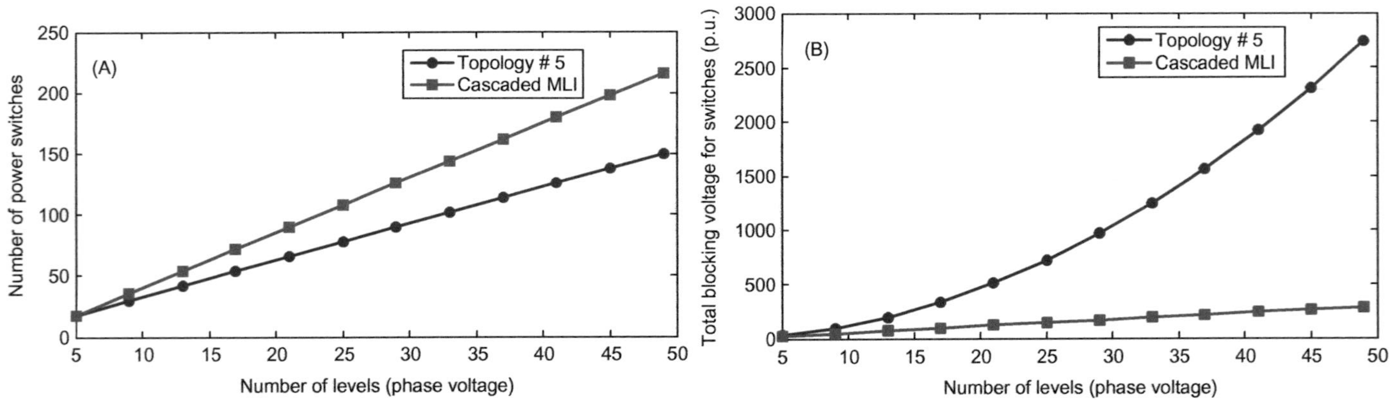

FIGURE 7.17 Comparison of cascaded MLI with "topology #5" (CBSC-MLI) on the basis of: (A) number of power switches required; and (B) total blocking voltage of switches.

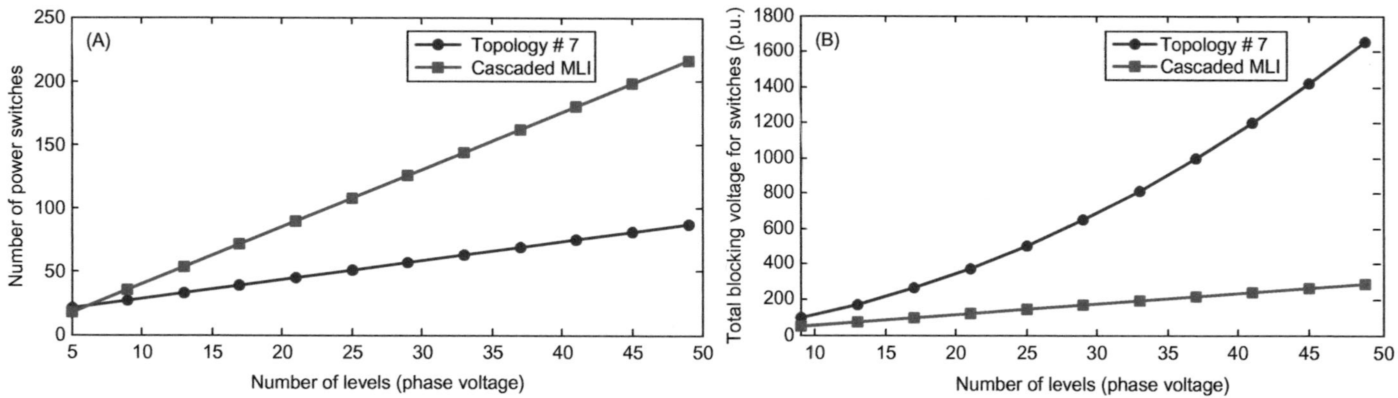

FIGURE 7.18 Comparison of cascaded MLI with "topology #7" (MLM-MLI) on the basis of: (A) number of power switches required; and (B) total blocking voltage of switches.

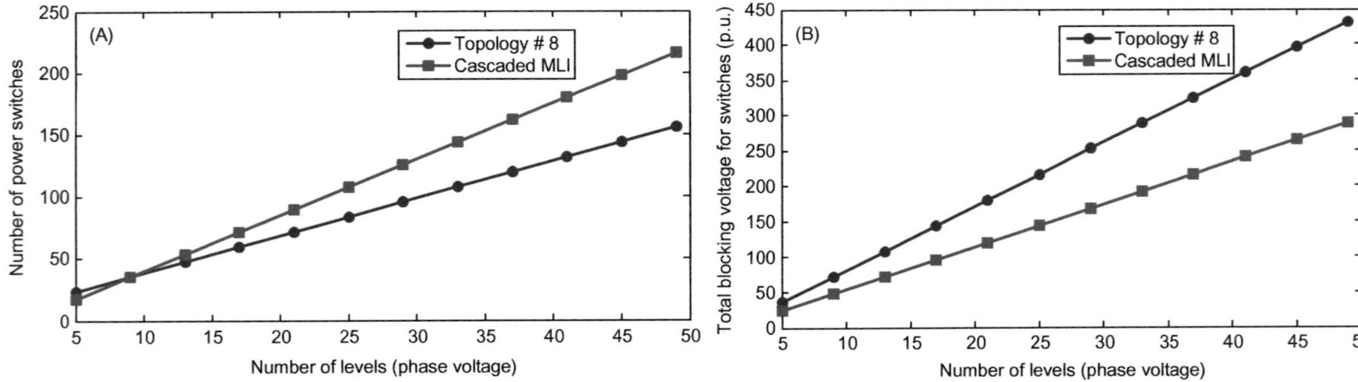

FIGURE 7.19 Comparison of cascaded MLI with "topology #8" (RV topology) on the basis of: (A) number of power switches required; and (B) total blocking voltage of switches.

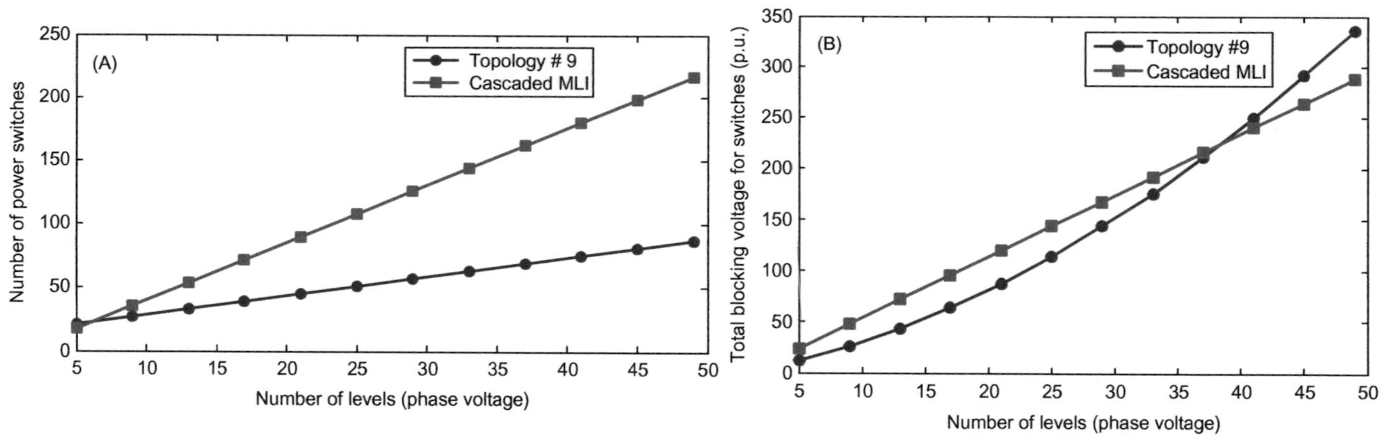

FIGURE 7.20 Comparison of cascaded MLI with "topology #9" (2SELG-MLI) on the basis of: (A) number of power switches required; and (B) total blocking voltage of switches.

REFERENCES

[1] J. Ebrahimi, E. Babaei, G.B. Gharehpetian, A new multilevel converter topology with reduced number of power electronic components, Ind Electronics IEEE Trans 59 (2) (2012) 655–667.

[2] M.F. Kangarlu, E. Babaei, A generalized cascaded multilevel inverter using series connection of submultilevel inverters, Power Electronics IEEE Trans 28 (2) (2013) 625–636.

Index

Note: Page numbers followed by "*f*" and "*t*" refer to figures and tables.

FSC
www.fsc.org
MIX
Papier | Fördert
gute Waldnutzung
FSC® C083411

Kolibri 360 GmbH
Ferdinand-Jühlke-Straße 7
99095 Erfurt, Deutschland
produktsicherheit@kolibri360.de